Diagnosis of the Nitrogen Status in Crops

Springer
Berlin
Heidelberg
New York
Barcelona
Budapest
Hong Kong
London
Milan
Paris
Santa Clara
Singapore
Tokyo

Gilles Lemaire (Ed.)

Diagnosis of the Nitrogen Status in Crops

With 119 Figures and 17 Tables

Springer

Dr. GILLES LEMAIRE
INRA
Station d'Ecophysiologie
des Plantes Fourragères
86600 Lusignan
France

ISBN-13: 978-3-642-64506-8 e-ISBN-13: 978-3-642-60684-7
DOI: 10.1007/978-3-642-60684-7

Library of Congress Cataloging-in-Publication Data. Diagnosis of the nitrogen status in crops/Gilles Lemaire (ed.). p. cm. Includes bibliographical references (p.) and index.
1. Nitrogen fertilizers. 2. Crops and nitrogen. 3. Plants. Effect of nitrogen on. I. Lemaire, Gilles. 1945- . S651.D53 1997 631.8′4 — DC21 96-52379

Softcover reprint of the hardcover 1st edition 1997

Cover design: Design & Production GmbH, Heidelberg

Typesetting: Best-set Typesetter Ltd., Hong Kong

SPIN: 10501692 31/3137/SPS – 5 4 3 2 1 0 – Printed on acid-free paper

Foreword

This volume introduces a new concept in plant nutrition and demonstrates how it can be applied in practice to improve N nutrition of crops. The importance of the topic is well illustrated by the fact that yields of man's main food, cereals, have increased, in many countries, roughly in proportion to average rates of N-fertilizer application. Yet much of this nitrogen does not enter the crops but is converted to nitrate, thus polluting the environment. Of course, in addition to fertilizers, biological nitrogen fixation and organic manures supply much nitrogen, however, regardless of the source, there is a need to match supply with crop demand.

Crop demand is the product of plant mass multiplied by critical N% (the minimum needed for maximum growth rate); the value of which usually declines with increasing plant mass. Much effort in the past was devoted to measuring critical N%. However, no set of values seems to have gained widespread acceptance, possibly because of divergence between results of different workers. Even so, important and widely applicable conclusions are emerging. One is the discovery, made independently in Lusignan and in Wellesbourne, of a remarkable similarity in the way the critical N% of some quite different crops declines with increase in plant mass per unit area. Others include the influence on the critical N% of the relative proportions of structural and metabolic tissues, of light intensity and its attenuation down the canopy, and of the formation of storage tissues. The authors have integrated this information into a comprehensive theory for the decline in critical N% as crops grow. The theory is consistent with the near proportional relationships that are often obtained between relative growth rate and N% in the plant dry matter and, even more notably, with the 3/2 self-thinning rule for plant populations in plant ecology.

A great merit of the theory is that it is encapsulated in a simple equation with two easily measurable coefficients. It has been extended to provide a nitrogen nutrition index that can be used to assess the nitrogen status of crop in terms of its N% and critical N% calculated from the equation and plant mass per unit area.

Failure to take proper account of plant mass per unit area has, in the past, led to much misinterpretation of field experimental data obtained in both temperate and tropical climates. Numerous examples are given of how reinterpretation of the data using the nitrogen nutrition index has led to discoveries. They cover disparate topics, including breeding of high-N lucerne cultivars, discriminating between the direct effects of plant water stress on growth and the indirect effects associated with reducing the N supply to roots, and elucidating the causes of differences in performance between mixed populations and monocultures of herbage. The studies were made at the field level and thus should lead rapidly to improvements in commercial practice.

The research can also be immediately applied to optimize the use of fertilizers. For instance, it provides relationships between critical N% and plant mass that can be incorporated directly into computer models that calculate the day-to-day changes in N response. As such models are now being used commercially by consultants, advisers and farmers, they could have a substantial impact on arable agriculture. Nevertheless, it is true that the decision-making process on fertilizer use on farms is complex and it is not clear how fertilizer practice will develop in the future. Inclusion of a discussion of this topic is a very welcome feature of the book, especially as it seldom gets the attention it deserves. Existing methods for predicting fertilizer requirements and, indeed, future possibilities, including those involving noninvasive radiometry, are also disscussed. The value of the nitrogen nutrition index is discussed in the light of this information and it is clear that it could be used in several different ways to improve N-fertilizer practice.

The book will be of great interest to all concerned with the fundamental aspects of plant nutrition as well as to agriculturalists, horticulturists and ecologists.

H.R.I. Wellesbourne, UK Prof. D.J. Greenwood

Contents

List of Contributors

AUBRY, C.
INRA-INA PG Unité de Recherche sur les Systèmes Agraires
et la Développement, 78850 Thiverval-Grignon, France

BARET, F.
INRA Station de Bioclimatologie, Domaine Saint Paul,
Site Agroparc, 84914 Avignon Cédex 9, France

CRUZ, P.
INRA Station d'Agronomie, BP 27,
31326 Castanet-Tolosan Cédex, France

DORE, T.
INRA-INA PG Laboratoire d'Agronomie,
78850 Thiverval-Grignon, France

DUCHENNE, T.
Agro-Transfert, Domaine de Brunehaut, 80200 Estrées-Mons,
France

DURU, M.
INRA Station d'Agronomie, Auzeville BP 27,
31326 Castanet-Tolosan, France

FOURTY, TH.
INRA Station de Bioclimatologie Domaine Saint Paul,
Site Agroparc, 84914 Avignon Cédex 9, France

GASTAL, F.
INRA Station d'Ecophysiologie des Plantes Fourragères,
86600 Lusignan, France

GRINDLAY, D.
INRA Station d'Ecophysiologie des Plantes Fourragères,
86600 Lusignan, France

Jeuffroy, M.H.
INRA-INA PG Laboratoire d'Agronomie,
78850 Thiverval-Grignon, France

Justes, E.
INRA Unité d'Agronomie Centre de Recherche Agronomique
2 Esplanade Roland Garros, BP 224, 51686 Reims Cédex 2, France

Le Bail, M.
INRA-INA PG Unité de Recherche sur les Systèmes Agraires
et la Développement, 78850 Thiverval-Grignon, France

Lemaire, G.
INRA Station d'Ecophysiologie des Plantes Fourragères,
86600 Lusignan, France

Machet, J.M.
INRA Station d'Agronomie, rue Fernand Christ,
02007 Laon Cédex, France

Martin, M.
ITCF-ITPT, Domaine de Brunehaut, 80200 Estrées-Mons,
France

Mary, B.
INRA Unité d'Agronomie, rue Fernand Christ,
02007 Laon Cédex, France

Meynard, J.M.
INRA-INA PG Station d'Agronomie,
78850 Thiverval-Grignon, France

Ney, B.
INRA-INA PG Laboratoire d'Agronomie,
78850 Thiverval-Grignon, France

Plénet, D.
INRA Unité d'Agronomie, 71, av. Edouard Bourleaux,
BP 81, 33883 Villenave D'Ornon Cédex, France

Sagan, M.
INRA Station de Génétique et Amélioration des Plantes,
Domaine d'Epoisses, 21110 Bretenières, France

Soussana, J.F.
INRA Station d'Agronomie, 12 Avenue du Brézet,
63039 Clermont-Ferrand Cédex 02, France

PART I

On the Critical N Concentration in Agricultural Crops

CHAPTER 1

N Uptake and Distribution in Plant Canopies

G. LEMAIRE and F. GASTAL

Contents

1.1 Introduction

Nitrogen (N) is often considered to be the most important limiting factor, after water deficiency, for biomass production in natural ecosystems. In arable and forage cropping, N fertilization practices can provide a sufficient N supply for plants to achieve the potential growth allowed by the amount of energy intercepted by the crop. However, to ensure that this potential yield is reached, the N inputs are often higher than the minimum required for maximum crop growth: this is particularly true because N fertilizers are relatively cheap compared to the expected economic benefits from a maximized crop yield.

At the present time, farmers must take into account more and more the environmental consequences of their practices, and hence must adjust N inputs according to the N requirement of the crops corresponding to the target yield. Due to new economic constraints, the target yield does not correspond exactly to the potential yield permitted by climatic conditions, and the optimization of the economic benefits to the farmer could require a lower yield than the maximum. Thus,

INRA Station d'Ecophysiologie des Plantes Fourragères, 86600 Lusignan, France

Diagnosis of the Nitrogen Status in Crops
G. Lemaire (ed)

from practices where a large amount of N fertilizer was applied to be sure of avoiding crop N deficiency in any situation, farmers have to move to a new N fertilization strategy which implies a determination of both the potential yield of any crop in a given climatic condition and the corresponding minimum crop N requirement necessary to achieve it. It is also necessary to estimate the restriction in N supply necessary to maintain the crop yield at just the optimum in economic terms. Despite the recent large improvement in modeling of crop growth, the prediction of both crop yield and minimum N fertilizer requirement for maximum growth is difficult at the farm level because of the unpredictable climatic parameters which determine plant growth and soil N availability, and because of our lack of knowledge about the basic mechanisms which govern N cycling. Faced with this variability, and having to anticipate N fertilization decisions, farmers have tended to overestimate the N fertilizer requirements of their crops. The possibility of diagnosis of the N status of any crop during its development is a valuable tool for making decisions on the need to apply, or not, a supplemental amount of N.

The objective of this first chapter is to develop the basic theory to support the concept of critical crop %N. This concept states that, with a given crop dry matter, a certain plant N concentration in the dry matter (N%) is needed to obtain the maximum instantaneous crop growth rate. According to this concept, it would be possible to use the critical plant N% to develop a tool for diagnosis of the N status of crops; but before such a diagnostic method can be put forward, the concept of critical N% has to be developed at both the individual plant and the canopy levels, taking into account the effect of competition for light inside a dense crop on N distribution between individual plants and between organs within the plant in relation to canopy structure and plant density. For this purpose, crops have to be considered not only as a global entity which exchanges mass and energy with the external environment, but also as a population of individual biological units which interact among themselves under a constraint of limited energy resources. Because our purpose is to develop an approach which is largely species-independent, it is necessary to identify as clearly as possible which parameters are dependent on the functioning of individual biological units (i.e., genetically dependent) and which parameters are dependent on the crop structure itself (i.e., ecophysiologically dependent), resulting from the overall interaction between plant population and microenvironment.

1.2 Maximum Growth Rate and N Accumulation in Crops; the Concept of Critical Plant N%

1.2.1 Relationships Between Dynamics of N Uptake and Dynamics of Biomass Accumulation for Different Crops

Even when there is an ample supply of N, the N concentration in plants within dense canopies declines as they grow. This phenomenon has usually been interpreted as the result of plant aging and has often been related simply to time, leading to large differences between species in a given situation and between different growing conditions for a given genotype. Lemaire and Salette (1984a,b) demonstrated that for grasses growing with nonlimiting N supply, the N concentration of a sward can be related to the dry matter accumulation by a simple equation:

$$N\% = a(W)^{-b}, \quad (1)$$

where W is the weight of aboveground biomass in $t\,ha^{-1}$ and N% is the N concentration as a percentage of dry matter.

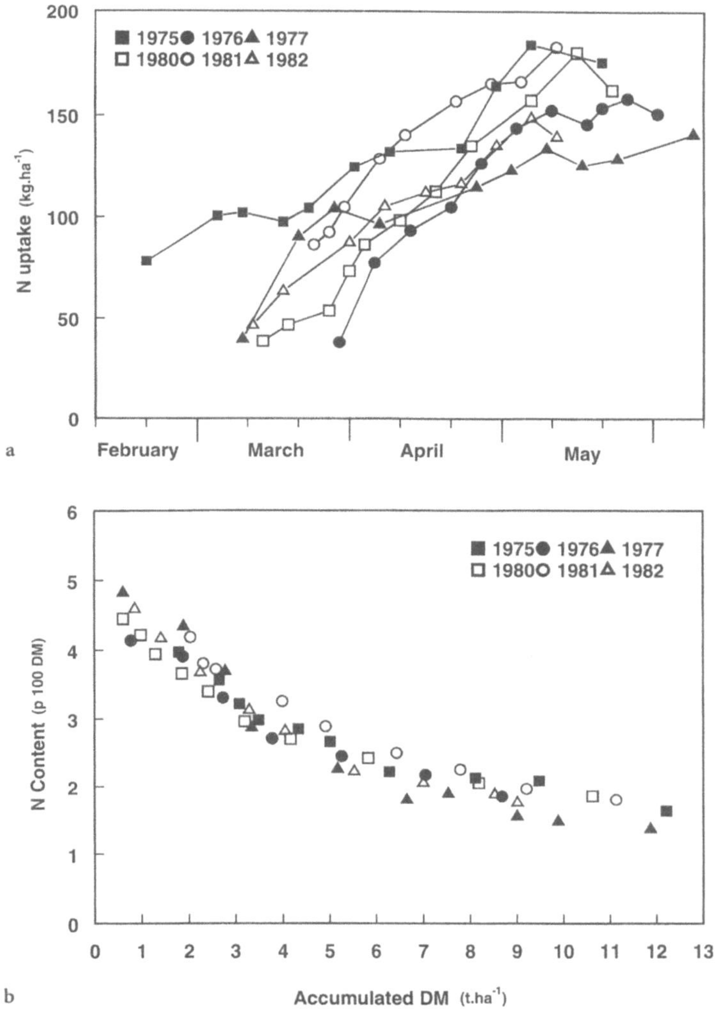

Fig. 1.1. **a** Dynamics of N uptake of tall fescue swards in spring under nonlimiting N supply, **b** Pattern of decrease in plant N% with increasing biomass for tall fescue swards during spring growth

This relationship can be transformed by multiplying the two sides of Eq. (1) by W, allowing an expression of the dynamics of accumulation of N by the sward in relation to the dynamics of accumulation of dry matter in the aboveground biomass:

$$N_{upt} = 10a(W)^{1-b}, \tag{2}$$

where N_{upt} is N accumulation expressed in $kgN\,ha^{-1}$.

a

b

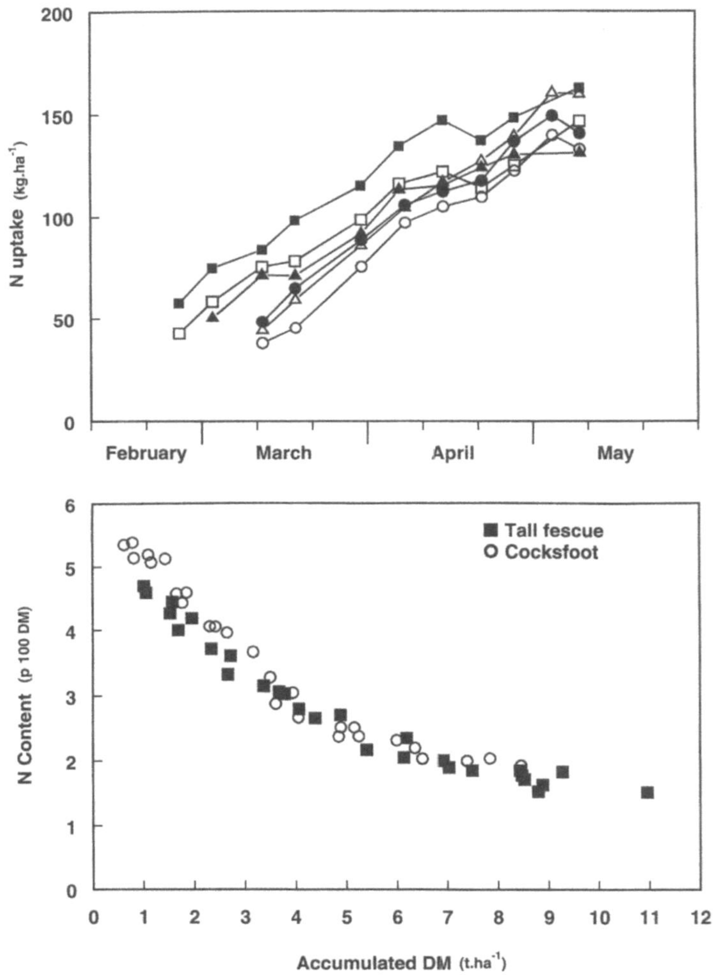

Fig. 1.2. **a** Dynamics of N uptake during spring growth for six genotypes of tall fescue and cocksfoot. tall fescue: M77 (■); Clarine (▲); Ludelle (●); cocksfoot: Cambria (□); Lubryde (△); Lucifer (△). **b** Decrease in plant N% with increasing sward biomass for tall fescue (■) and cocksfoot (△)

The coefficient a in Eq. (1) represents the plant N% for an aboveground biomass of $1\,t\,ha^{-1}$; in consequence, the coefficient 10a of Eq. (2) corresponds to the quantity of N taken up by the sward when the aboveground biomass reaches $1\,t\,ha^{-1}$. The coefficient b of Eq. (1) characterizes the pattern of decrease of N% during regrowth. If $b = 0$, N% remains constant during the regrowth; if $b = 1$, there is an increase in aboveground biomass without any increase in N accumulation. The coefficient $1 - b$ of Eq. (2) represents the ratio of the relative rate of N accumulation (dN/Ndt) to the relative rate of shoot biomass accumulation (dW/Wdt). A value of $1 - b < 1$ indicates that, even with nonlimiting N supply, N accumulates at a lower relative rate than does dry matter during sward regrowth, leading to a decrease in plant N concentration.

Figure 1.1a represents the quantities of N taken up by a tall fescue (*Festuca arundinacea* Schreb.) sward in different years during a spring regrowth after a nonlimiting N supply at the end of winter. Large differences can be observed between years, but the pattern of plant N% decrease with increasing sward biomass is similar for the different years (Fig. 1.1b). Thus, differences in N uptake between years can be attributed solely to differences in sward growth dynamics as a consequence of climatic conditions (Lemaire and Salette 1984a).

Figure 1.2 shows similar results for different genotypes of tall fescue (*Festuca arundinacea* Schreb.) and cocksfoot (*Dactylis glomerata* L.). This indicates that the differences in plant N% are related not only to differences in dry matter accumulation due to environmental conditions other than soil N availability, but also to genotypical growth capacities under given environmental conditions. From these data, Lemaire and Salette (1984a) proposed a stable value for coefficients a and b across sites, years and cultivars, for swards receiving sufficient N supply to ensure that growth is not restricted by N availability in soil:

$$N\% = 4.8(W)^{-0.32} \qquad \text{for Eq. (1)}$$

and:

$$N_{upt} = 48(W)^{0.68}. \qquad \text{for Eq. (2)}$$

The generalization of these coefficients to nongrass species was tested on *Medicago sativa* L. (lucerne) stands in different sites and for different regrowth periods by Lemaire et al. (1985). Table 1.1 shows that the values of coefficients a and b proposed for well-fertilized grasses are similar to those obtained for lucerne in the cases where N fixation was probably not limiting growth.

A similar approach has been taken for various C_4 species such as *Sorghum sudanense* (Sudan grass), *Zea mays* (maize), and *Setaria anceps*, growing in either tropical or temperate situations with sufficient N supply. For this data set, when restricted to the period of vegetative growth, i.e., before the accumulation of storage products in grains, Lemaire and Chartier (1992) proposed a single value for coefficients of the Eq. (1):

Table 1.1. Comparison of coefficients 10a and (1-b) of Eq. (1) for different regrowths of lucerne crops. (Lemaire et al. 1985)

	10a	1-b	R^2
Spring growth	55	0.64	0.978
First regrowth	48	0.69	0.996
Second regrowth	46	0.67	0.970
Third regrowth	47	0.71	0.968

$$N\% = 3.6(W)^{-0.34}.$$

An analysis of the value of coefficients a and b for a larger data set of C_3 species (tall fescue, lucerne, potato, wheat, rape, beans, cabbage) and C_4 species (maize, sorghum, setaria) was carried out by Greenwood et al. (1990). These authors concluded that, when data corresponding to periods of storage material accumulation in plants were excluded, i.e., during vegetative growth period before grain (or tuber) filling, a unique value for coefficient b of 0.34 could be proposed for both C_3 and C_4 crops.

The difference in the value of coefficient a between C_3 and C_4 species, respectively 4.8 and 3.6, reflects the difference in the metabolic pathways for CO_2 assimilation and the associated differences in leaf anatomy (Brown 1978, 1985; Field and Mooney 1986). The value of 4.8 found for C_3 crops corresponds to the value for maximum growth rate determined for young lettuce plants by Burns (1992), and is not very different from the value of 4.3 proposed as the optimum N% in leaves of seedlings of *Betula pendula* L. by Ingestad (1979). The value of 3.6 found for C_4 crops corresponds with the range of optimum N concentrations (3.5–3.8%) reported for leaves of maize (Hanway 1962; Voss et al. 1970). The difference in the value of coefficient a for the two types of species indicates that, under nonlimiting N supplies, C_4 crops require only 75% of the N required by C_3 crops for the same biomass production, as a consequence of Eq. (2). Since the increase in dry weight per unit of intercepted photosynthetic active radiation, that is the radiation use efficiency (Monteith 1977), had been estimated to be 1.9 g MJ^{-1} for various C_3 crops and 2.5 g MJ^{-1} for various C_4 crops (Gosse et al. 1986), i.e., approximately 25% lower for C_3 than for C_4, the rate of N uptake (or protein produced) per unit of intercepted radiation is approximately the same for the two types of crops when compared at similar biomass (Greenwood et al. 1990).

1.2.2
Critical N% for Maximum Growth Rate

In any situation and at any given time, critical plant N% can be defined as the minimum N concentration in plants needed for maximum growth rate of the crop (Ulrich 1952). Similarly, it is possible to define a critical N uptake as the minimum quantity of N necessary to achieve maximum aboveground biomass at a given time. Such a definition implies that N% and W in Eqs. (1) and (2) correspond to conditions with the minimum N supply for maximum growth during all the growth period. This condition can be effectively realized in controlled conditions using the addition rate technique developed by Ingestad and Lund (1979). In the field, this condition cannot be maintained as precisely as is theoretically possible

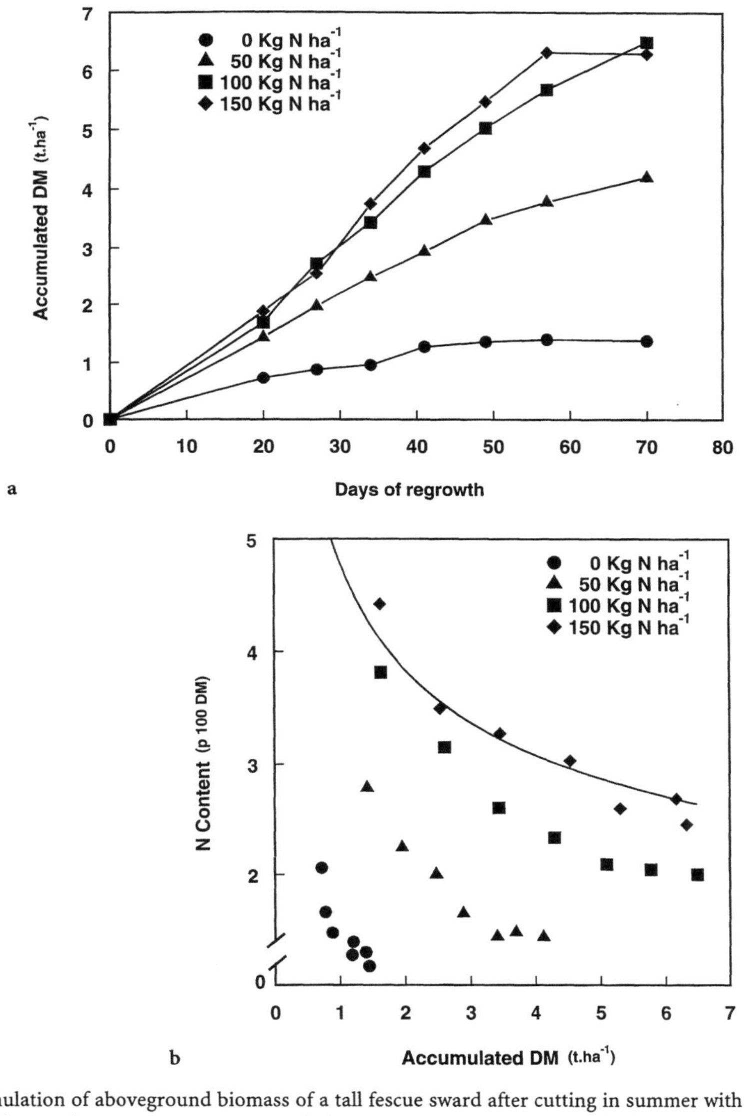

Fig. 1.3. **a** Accumulation of aboveground biomass of a tall fescue sward after cutting in summer with different levels of N application. **b** Time course of plant N% in relation to sward biomass during regrowth for each level of N supply. The *line* corresponds to Eq. (1): $N\% = 4.8(W)^{-0.32}$

during the growth period. Even with adequate N supply, crops can fluctuate from situations of "luxury" N consumption to situations of suboptimal N nutrition according to changes in soil N availability. Thus, the determination of critical N% from data obtained in field conditions requires careful procedures.

The fitted values for a and b obtained from the data sets used in Section 1.2.1 correspond to conditions where N supply was probably large enough to avoid limitation of growth by N deficiency but was also not excessive, to avoid luxury consumption. All the data were obtained in multiharvest experiments where at least three levels of N supply were applied. For each experiment and each harvest date the highest dry weight obtained with any level of fertilizer, and the corresponding %N, were identified and included in the data base. When the highest yields were obtained with each of two or more levels of fertilizer, the data obtained with the lower level were selected.

For the tall fescue experiment (Fig. 1.3), for the first two harvest dates, the N supply of 100 kg ha^{-1} appeared to be sufficient to achieve maximum growth rate and the two corresponding points can be considered as near critical N%. For the four following harvest dates, the N supply of 150 kg ha^{-1} allowed a small but significative increase in growth rate and these data points have been considered as corresponding nearly to critical N%. The seventh harvest date has to be omitted because of lodging. For the sorghum experiment (Fig. 1.4) we observed neither an increase in growth rate at any period in response to the three levels of N supply, nor any substantial increase in plant N%. Thus all the data points corresponding to the lowest level of N supply have been considered as close to critical N%. This procedure allows only the selection of data points corresponding to near-optimum N nutrition conditions, but is not sufficiently precise because some data points corresponding to luxury consumption of N, and also some others corresponding to suboptimal N nutrition, can be introduced in the data set. This difficulty in selecting accurate data points corresponding exactly to optimal N nutrition becomes particularly important for experiments with only three levels of N supply. Despite this problem, we can argue that for a large data base including several experiments, a relatively equal proportion of data points should correspond to either a small degree of N luxury consumption or small degree of N deficiency, and therefore the average fitted equation between N% and W should correspond approximately to the critical %N.

Justes et al. (1994) proposed a more precise statistical method for the determination of data points corresponding to the critical N%. These authors used several experiments on wheat with five or six levels of N supply. At each measurement date, the aboveground biomass (W) was compared in the different N treatments with a Student's two-tail t test in order to separate a group of points of similar W from points where W increased with increasing N supply. The variation of N% vs. W was described as a bilinear relationship composed of: (1) an oblique regression line corresponding to the increase in W with the increase in N%; (2) a vertical line corresponding to an increase of plant N% without any increase in W. The critical N% is therefore determined by the ordinate of the intersection point of the two lines. This value and the corresponding value of W can be then introduced in a

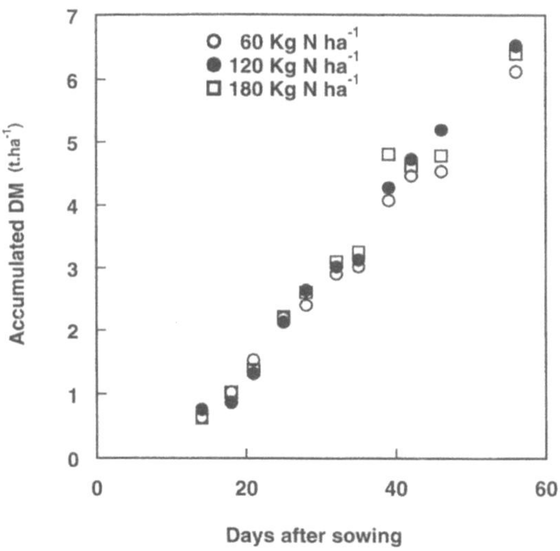

a

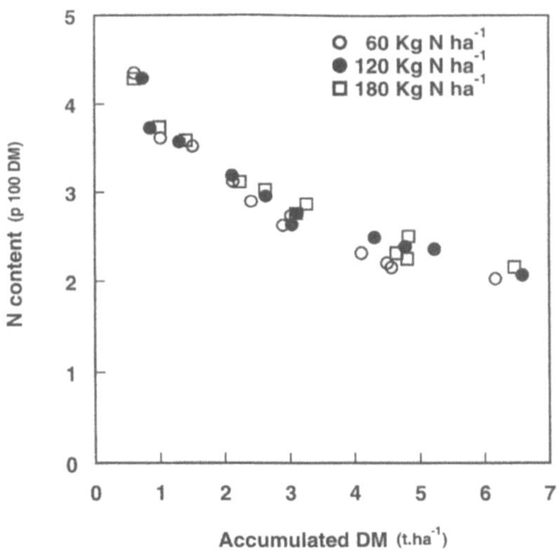

b

Fig. 1.4. **a** Aboveground biomass of a forage sorghum crop (*Sorghum sudanense* cv. Piper) grown with different N fertilizer levels at different times after the first cut (day 220). **b** Relationships between plant N% and crop mass for the different fertilizer N levels

data base for determining the coefficient a and b of the Eq. (1). The results obtained by Justes et al. (1994) for wheat with this procedure gave a relationship close to that proposed by Greenwood et al. (1990) for C_3 crops. Plenet (1995), also using this procedure, determined values for coefficients a and b for maize before grain filling similar to that proposed by Lemaire and Chartier (1992) for C_4 crops.

Thus, there is strong evidence that critical plant N% declines in a similar pattern as the crop grows for a large range of species of the same metabolic group.

The fractional decline in N% per unit of aboveground biomass increase [coefficient b of Eq. (1)] appears to vary only between 0.32 and 0.44 for a large number of species, which leads to variation of the ratio between relative N accumulation rate and relative growth rate between 0.68 and 0.56 during the period of vegetative growth for all the species studies [coefficient $1-b$ of Eq. (2)]. The relative stability of coefficients a and b could confer to Eqs. (1) and (2) a status of "law". Because of the high correlation between the values of a and b, these relationships can be considered as largely species-independent in a range of cultivated crops, if storage organs are excluded. Nevertheless, such an empirical relationship must be explained in terms of ecophysiological processes before being generalized to a wider range of situations and species, and before it is used as a reference for the critical level of N nutrition of crops in agronomical studies.

1.3 Ecophysiological Basis for the Decrease in Critical Plant N% with Plant Mass

The decrease in plant N% with increase in crop mass indicates that, as a crop grows, N accumulates at a progressively lower relative rate than carbon (C). If we admit that the C concentration of biomass is approximately constant during period of vegetative growth, i.e., when the crop accumulates only leaf and stem tissues, then Eq. (2) can be expressed as the result of an allometric relationship between accumulation of N and accumulation of C in aboveground biomass:

$$N = k(C)^{1-b}. \quad (3)$$

The coefficient k represents the N:C ratio of the crop when the first unit of C is accumulated. The coefficient $1-b$ represents the ratio between the relative rates of N and C accumulation:

$$\frac{dN}{Ndt} = (1-b)\frac{dC}{Cdt} \quad (4)$$

or:

$$\frac{dN}{dC} = (1-b)\frac{N}{C}. \quad (5)$$

Equation (5) indicates that at any time in the growth period, the N:C ratio of biomass newly synthesized during period dt is less than the N:C ratio of the biomass already accumulated. This implies that as the plant grows, there is a decline in the minimum quantity of N required per unit of assimilated C to give the maximum growth rate. Thus the nitrogen productivity, i.e., the quantity of C assimilated per unit of absorbed N, increases with crop growth. The concept of nitrogen productivity was introduced by Agren (1985). This author found that the

N productivity of young seedlings remained constant when the plants were maintained at constant N nutrition levels. The constancy of N productivity was obtained only for periods where relative growth rate (RGR) could be considered as constant, i.e., for young plants in nearly exponential growth. Under these conditions, and for different levels of N supply, RGR has been found to be linearly related to plant N% (Ingestad 1979; Ericksson 1981; Agren 1985). However, these conditions of constant relative growth rate are far from those of crop growth in the field, where RGR declines as the crop grows, even in constant climatic conditions and when supplies of nutrients and water meet the crop demand. Thus it is necessary to develop a theory which accounts for the increase of N productivity and the corresponding decrease in plant N% in crops out of the juvenile phase. However, this theory must remain consistent with Agren's theory for the early stage of crop development. We have to note here that for very low crop mass ($W \rightarrow 0$) Eq. (1) gives unrealistic values for plant N% and cannot be used for the early stage of growth, i.e., before a crop aboveground biomass of approximately $1\,t\,ha^{-1}$.

1.3.1 A General Model of Plant N% Decline with Increasing Crop Mass

In the early stage of growth, even in dense crops, individual plants can be considered as isolated, and their growth is often approximated as being exponential, i.e., with a constant RGR. Rapidly, as the canopy develops, interference between individual plants occurs by mutual shading and RGR declines. Caloin and Yu (1984) developed a conceptual model linking the decrease in RGR with the decline in plant N%. The main assumption of this model is that N in the plant is associated either with (1) metabolic activity (mainly photosynthesis), or (2) structural tissues. If the plant mass per unit area associated with (1), i.e., metabolic plant mass, is W_m and that associated with (2), i.e., structural plant mass, is W_s then:

$$W = W_m + W_s, \tag{6}$$

where W is the total plant mass per unit area. The absolute growth rate must, by definition, be proportional to W_m, the only active component of plant. Thus:

$$\frac{dW}{dt} = kW_m, \tag{7}$$

where k is the growth coefficient which is constant for a given environmental condition. Rearranging Eq. (7) gives a relationship between the fraction of the metabolic component of plant mass and the RGR:

$$\frac{W_m}{W} = \frac{1}{k}\frac{dW}{Wdt}. \tag{8}$$

If N, N_m, and N_s are the N concentrations in the whole plant dry matter (W), and in the two components W_m and W_s respectively, then:

$$N = \frac{N_m W_m + N_s W_s}{W}. \tag{9}$$

Using Eqs. (6), (8), and (9) gives:

$$N = \frac{1}{k}(N_m - N_s)\frac{dW}{Wdt} + N_s. \tag{10}$$

If we assume that the N concentration of each plant component is constant with time, Eq. (10) predicts a linear relationship between plant N% and crop RGR. For the early stage of growth, under conditions of nearly exponential growth with an approximately constant RGR, plant N% should remains constant, as proposed by Agren (1985). The subsequent decline in RGR with increasing crop mass is accompanied by a decrease in plant N%, which, according to the constancy of N_m and N_s, results solely from the decline in the ratio W_m/W_s as crop mass increases. Thus, as the canopy develops, plants contain an increasing proportion of structural tissues that contain less N. The W_m component corresponds mainly to photosynthetic machinery, and the associated N concentration, N_m, is high compared to that of the structural tissues, N_s.

Greenwood et al. (1991) showed that Eq. (10) holds for different species and for different levels of N supply. Smoothing was necessary to estimate RGR from successive harvest data and these authors used the following growth model:

$$\frac{dW}{dt} = K_x \frac{W}{x + W}, \tag{11}$$

giving:

$$RGR = \frac{dW}{Wdt} = K_x\left(\frac{1}{x + W}\right), \tag{12}$$

where x is a coefficient which represents the value of W for which RGR becomes half of its initial value and K_x is the growth coefficient. K_x and x are highly correlated and cannot be estimated independently. Comparison between Eqs. (8) and (12) allows us to establish a relationship between $1/(x + W)$ and W_m/W, i.e., the fractional content of metabolic component in total plant mass:

$$\frac{W_m}{W} = \frac{K_x}{k}\left(\frac{1}{x + W}\right). \tag{13}$$

If we assume that $W_m/W = 1$ when $W = 0$, then $K_x/k = x$, and substituting RGR from Eq. (12) into Eq. (10) gives:

$$N = (N_m - N_s)\frac{x}{x + W} + N_s. \tag{14}$$

This hyperbolic formulation of the decline of plant N% with crop biomass allows us to account for a minimal N concentration in the shoot which corresponds to the structural N pool. Plant N% tends to N_s for very high values of crop mass. Similarly, N_S represents the minimum N% that can occur in plant, i.e., that at which RGR becomes zero. At the initial growth stage, that is for low values of W, plant N% is approximately equal to N_m: at this stage, plant mass is entirely composed of metabolic material. Thus, Eq. (13) provides a more realistic model than Eq. (1), which tends to overestimate plant N% for low crop biomass. Such an equation can be used for modeling plant N% decline with increasing crop mass.

The use of data of Justes et al. (1994) on wheat crops and Plénet (1995) on maize crops allowed the determination of the optimized values of N_s, N_m, and x (Table 1.2). A common value of 0.8% for N_s can be used for wheat and maize crops. This value is similar to that determined by Greenwood et al. (1991) on tall fescue. These authors showed that, according to the basic assumption of the Caloin and Yu (1984) model, this value remains constant during the growth of the crop. The optimized values of N_m correspond to an estimation of the N% of the plant mass directly associated with growth, that is the definition of W_m in Caloin and Yu's model. This metabolic component of the plant is only a theoretical concept and is not precisely identified as a morphological or histological compartment. Nevertheless, we can assume that W_m mainly represents the mass of the active photosynthetic tissues, and therefore N_m represents their concentration of N. Thus, the estimated values of 6.5 and 4.8% for C_3 and C_4 plants reflect the difference in Rubisco concentration in photosynthetic paremchyma between the two types of species. According to Eq. (13), and if we allow that $K_x/k=x$, then:

Table 1.2. Optimization of the coefficient Ns, Nm, and x of Eq. (14), using data of Justes et al. (1994) for wheat and Plénet (1995) for maize. (B. Mary, unpubl.)

	Free optimization	Optimization with imposed values		
		Ns = 0.60	Ns = 0.40	x = 1
Maize				
Ns	0.82	0.60	0.40	0.94
Nm	4.81	4.13	3.77	6.05
x	1.75	2.99	4.40	1.00
Mean square error	0.16	0.24	0.41	0.21
Wheat				
Ns	0.77	0.60	0.40	1.21
Nm	6.47	6.11	5.80	9.46
x	2.48	3.02	3.69	1.00
Mean square error	0.19	0.20	0.21	0.26

$$\frac{W_m}{W} = \frac{x}{x+W}. \tag{13'}$$

The difference in the value of x for C_3 and C_4 crops, as shown in Table 1.2, suggests that at similar plant mass, C_4 crops contain a smaller proportion of metabolic tissues than C_3 crops. Thus, the difference in critical plant N% observed between the two groups of species when comparing at similar biomass appears to be the result of (1), a difference in N concentration of the metabolic compartment, and (2), a difference in the proportion of the structural compartment in the total plant biomass. This difference in the value of x between maize and wheat reflect the 40% greater radiation use efficiency (RUE) of C_4 plants compared to that of C_3 plants (Gosse et al. 1986), if we accept that the coefficient k of Eq. (7) represents the efficiency of light conversion into plant biomass under constant environmental conditions.

In conditions of N deficiency, Greenwood et al. (1991) showed that K_x was directly affected by plant N nutrition, leading to a linear decrease of K_x with decreasing plant N% resulting from N supply treatments; but, as shown in Fig. 1.5, K_x appears to be independent of the decrease in plant N% with time resulting solely from the plant growth. Thus, the decline in RGR for crops growing in suboptimum N supply conditions can be attributed to two distinct phenomena: (1) a decrease in RGR due to the increase in plant mass per se, leading to a parallel decrease in the corresponding value of critical N% (N_c); and (2) a decrease in RGR due to a proportional decrease of K_x (and therefore x) with the ratio $(N_a - N_s)/(N_c - N_s)$, with N_a being the actual plant N%. Figure 1.5 shows that the value of N_s corresponds to the minimum plant N% that can occur in plant, i.e., the N% at which growth just takes place, does not vary with time and seems independent of plant

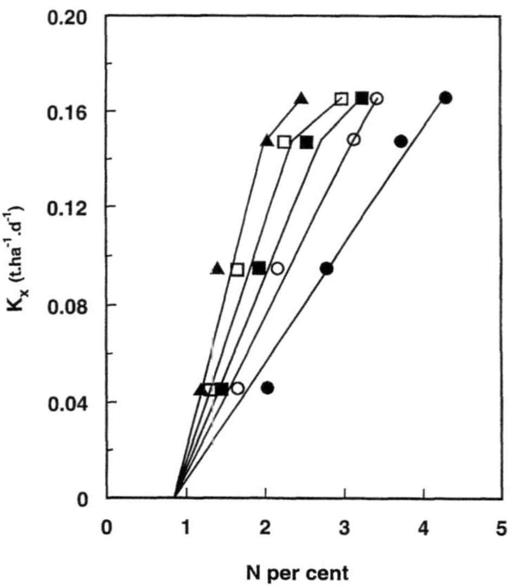

Fig. 1.5. Relationship between K_x [see Eq. (13)] and plant N% for tall fescue at different sampling dates during a regrowth period: 3 (●), 4 (○), 5 (■), 6 (□), and 7 (▲) weeks after cutting. (After Greenwood et al. 1991)

mass. According to Eq. (13′), the value (x/x + W) is equal to the ratio W_m/W, i.e., the proportion of metabolic component in the plant mass. Therefore, the decrease in the value of x under conditions of limiting N nutrition, as suggested by Greenwood et al. (1991), implies that the proportion of metabolic component in the plant decrease as the N deficiency increases.

The main difficulty in the application of such a model for plants growing in dense canopies is that the hypothesis of constancy of coefficient k of Eq. (7) supposes constant environmental conditions (temperature and radiation) for individual plants; but in dense crops the average illumination level of individual plants decreases as the canopy develops. Therefore, values of k should decline with increasing crop mass. Hence, in these conditions it becomes difficult to distinguish clearly between the two causes of the decline in RGR and plant N% during crop growth, the increasing competition for light between individual plants and (or) the increase in plant mass.

1.3.2 A Model of Plant N% Decline in Isolated Plants

Estimation of the effect of light competition on the plant N% decline in a dense canopy requires time-course analysis of plant N% during the growth of isolated plants in conditions of constant illumination. Growth of isolated plants is often considered exponential, i.e., with a constant relative growth rate (Charles-Edwards 1982; Philippot et al. 1991). Under these conditions, according to Ingestad (1979) and Agren (1985), plant N% should remain constant and independent of plant mass. Nevertheless, results obtained by Lemaire and Chartier (1992) on isolated sweet sorghum plants in field conditions, and by Allirand et al. (1992) on isolated lucerne plants in controlled conditions, show an unambiguous decline in plant N% with increase in plant mass. Caloin and Yu (1982) also obtained a decrease in RGR for isolated plants of *Dactylis glomerata* L. grown under controlled conditions. In order to explain their data, the latter authors developed the idea that the active part of plant mass, i.e., the metabolic component W_m, remains proportional to W^{α}:

$$W_m = c(W)^{\alpha}, \qquad \text{with } 0 < \alpha < 1 \tag{15}$$

where W is the total plant mass.

Using the logarithmic derivative of Eq. (15) gives:

$$\frac{dW_m}{W_m dt} = \alpha\left(\frac{dW}{W dt}\right). \tag{15′}$$

Thus, the coefficient α appears to be the ratio between the relative rate of increase in active plant mass (W_m) and the relative growth rate.

Substituting W_m from Eq. (7) into Eq. (15) gives:

$$\frac{dW}{dt} = kc(W)^{\alpha}. \tag{16}$$

The relative growth rate can be then deduced as follows:

$$RGR = \frac{dW}{Wdt} = kc(W)^{\alpha-1}. \tag{17}$$

Equation (16) postulates that plant growth rate is proportional to that component of its biomass which is active in the growth process. If $\alpha = 1$, Eq. (16) gives an exponential growth function and RGR remains constant, while values of $\alpha < 1$ lead to a decrease in RGR as a consequence of the decrease in the proportion of the active plant component with increasing plant mass:

$$\frac{W_m}{W} = c(W)^{\alpha-1}. \tag{18}$$

The substitution of the new expression of RGR of Eq. (16) in Eq. (10) gives a new formulation of plant N% decline with plant mass:

$$N = c(N_m - N_s)W^{\alpha-1} + N_s. \tag{19}$$

This expression indicates that plant N declines with plant mass with an asymptotic value corresponding to the N concentration of the structural component of plant. If N_s is very small, the formulation of Eq. (19) would not be very different from the empirical relationship of Eq. (1). It is important to note that the use of Eq. (18) for a very young stage of growth would lead to unrealistic values of $W_m/W > 1$ for values of $W < W_i = c^{1/1-\alpha}$. The term W_i can be interpreted as the initial plant mass when the plant is composed only of metabolic material. This should correspond to the plant mass at seedling emergence when autotrophic growth has just begun. Replacing c by $(W_i)^{1-\alpha}$ gives:

$$N = (N_m - N_s)\left(\frac{W_i}{W}\right)^{1-\alpha} + N_s. \tag{19'}$$

This equation allows us to describe the decline in plant N% with increase in plant mass from an initial plant mass W_i and an initial plant N% equal to N_m, down to an asymptotic value of plant N% equal to N_s.

The integration of Eq. (16) leads to a power relationship between plant mass W and time t:

$$W = \left[kc(1-\alpha)t + W_i^{1-\alpha}\right]^{\frac{1}{1-\alpha}}, \tag{20}$$

where W_i corresponds to initial plant mass for $t = 0$, i.e., at seedling emergence. The data of Lemaire and Chartier (1992) for sweet sorghum shows that the fitted value of α estimated by nonlinear regression is 0.85, with a value of W_i not different from zero. Such a value of α is reasonably close to the value of 0.79 calculated by Caloin and Yu (1982) for isolated plants of *Dactylis glomerata* L., and to the value of 0.82 found by Allirand et al. (1992) for isolated plants of *Medicago sativa* L. Nevertheless, in these last two references, plant mass W corresponded to total plant mass including roots, while for the sweet sorghum data only shoot mass was considered. Considering that root mass corresponded to about 10% of shoot mass for the two experiments on *Dactylis* and *Medicago*, we can conclude that for herbaceous species such as cocksfoot, lucerne, or sorghum, i.e., with the C_3 or C_4 photosynthetic pathways, a common value of 0.8 can be adopted for the coefficient α. Thus, according to Eq. (15′), the relative rate of increase in active plant mass ($dW_m/W_m dt$) is only about 80% of the relative growth rate, leading to a decrease in the ratio W_m/W with increasing plant mass.

It is necessary to note the importance of the conceptual assumption used in defining the metabolic component of plant mass, W_m. This concept is identical to that of metabolic weight developed by animal nutritionists for studying the energy requirements for animals of different sizes. A basic assumption for such a concept is that the exchange rates of a biological organism with the external environment, i.e., its metabolic activity, is proportional to the size of its exchange surfaces. For animals, an empirical allometric coefficient of 0.75 between surface and weight allows a metabolic weight $W^{0.75}$ to be calculated for comparisons of the performance of animals with different body weights. In fact, what is proposed in the model of plant growth developed by Caloin and Yu (1982) is simply to apply this concept of metabolic weight to plants. These authors developed the idea that the exponential growth model means that the active plant surface for exchange is proportional to plant mass. This implies $\alpha = 1$ in Eq. (15); but this hypothesis is verified only for two-dimensional expansion. For instance, the relationship between surface S and volume V of an object growing with the same relative rate in the three dimensions will be: $S = kV^{2/3}$. If we assume that the density remains constant, the relationship between the surface S and the weight W will be: $S = k'W^{2/3}$. To have proportionality between S and W, it is necessary to have expansion of the object in two dimensions only, i.e., an expanding area of constant thickness. Since the main active exchange surfaces in plants are leaves, the coefficient α of Eq. (15) should represent the allometric relationship between plant leaf area and plant mass. The value of α lower than 1 indicates that the ratio between plant leaf area and plant mass, which is usually named leaf area ratio (LAR), decreases as plant mass increases; but values of α greater than 2/3 indicate that the growth of an isolated plant does not appear to be isometric, i.e., having the same relative expansion rate in the three dimensions.

Optimization of light interception leads to plants having a bidimensional, rather than a tridimensional, expansion; but the mechanical constraints of bearing leaf surfaces in the optimal position for light interception obliges plants to invest a certain minimum dry matter in growth in the third dimension. Thus, as its leaf area develops, the plant must invest a greater proportion of its dry matter in supporting tissues, and therefore the value of $\alpha = 1$ necessary for exponential growth cannot be reached. The values for α of 0.80–0.85 found in different species seem to be the result of an optimization between light interception and mechanical constraints. This optimization can be reflected in ontogenic changes in leaf: stem ratio (L/S) or (and) in specific leaf weight (SLW).

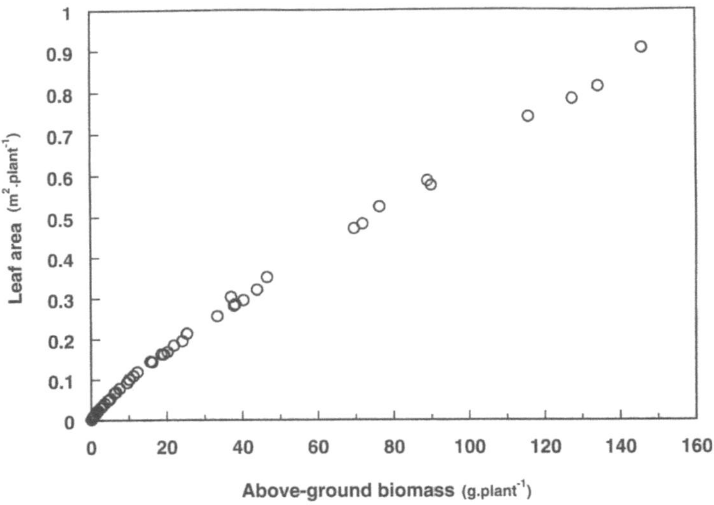

Fig. 1.6. Relationships between the dynamics of leaf area expansion and the dynamics of aboveground biomass accumulation for isolated plants of sweet sorghum. Each *point* represents an individual plant

Figure 1.6 shows that for isolated sweet sorghum plants (G. Lemaire and M. Chartier, unpubl. data), a narrow allometric relationship exists between the leaf area of the plant, S_L, and shoot biomass, W:

$$S_L = 142(W)^{0.83}.$$

This relationship is valid for both individual plants varying in size at a given date, and differences at various dates. The value of the allometric coefficient (0.83) is close to the estimated value of coefficient α: this contributes to validating the assumption that the metabolic component of plant mass is directly associated to leaf area. This equation allows the estimation of the decrease in leaf area ratio (LAR) with plant mass increase:

$$LAR = \frac{S_L}{W} = 142(W)^{-0.17}.$$

In the case of sorghum plants, this decrease in LAR is mainly associated with decrease in specific leaf area (SLA) as leaf weight (W_L) increases:

$$SLA cm^2\ g^{-1} = 189(W_L)^{-0.13}.$$

At the same time, the ratio of the biomass of leaf laminae to total shoot biomass (the leaf weight ratio, LWR) declines only slightly with increasing plant mass:

$$LWR = 72(W)^{-0.04}.$$

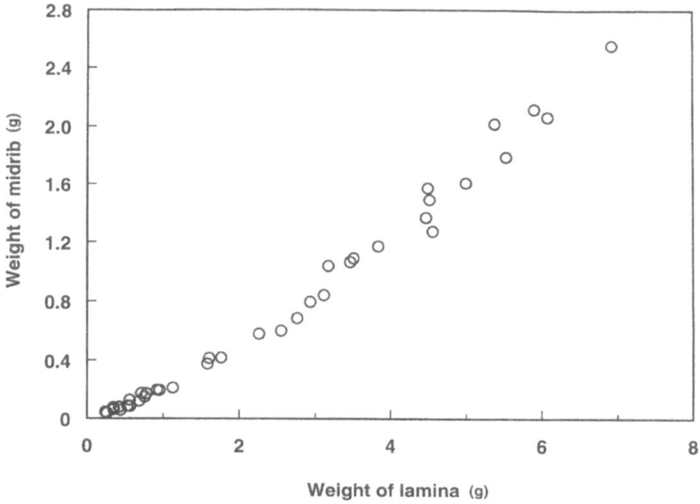

Fig. 1.7. Relationships between the weight of the midrib and the weight of the lamina for leaves of sweet sorghum plants growing in isolation. Each *point* represents an individual leaf

As illustrated in Fig. 1.7, the weight of the midrib of the lamina increases proportionately more than the total lamina weight. Thus, as the plant develops larger leaves, an increasing proportion of leaf structural tissues is necessary to maintain the lamina blade in erect position, which leads to the observed decrease in specific leaf area. Pons et al. (1993) noted that in *Carex acutiformis* and *Brachipodium pinnatum*, the increase in specific leaf weight (SLW) from the tip to the base of lamina was mainly associated with an increasing proportion of leaf tissues other than mesophyll, such as cuticule, vascular bundles, and sclerenchyma. Hence, as the individual plants accumulate successive leaves of increasing size, the accumulation of structural component of plant biomass is more rapid than the accumulation of metabolic component associated with leaf area. This leads to a shoot growth rate proportional to $W^{0.83}$ and a decline in relative growth rate proportional to $W^{-0.17}$, as described by Eqs. (16) and (17).

The decline in RGR of isolated plants is accompanied by a decline in plant N% as shown in Fig. 1.8. The use of Eq. (1) allows a fit to be made to the experimental data:

$$N\% = 5.06(W)^{-0.106} \qquad R^2 = 0.866.$$

Fitting the data to Eq. (19) with N_s = 0.8 allows us to increase only slightly the variance explained by the model:

$$N\% = 4.27(W)^{-0.128} + 0.8 \qquad R^2 = 0.873.$$

The small improvement of the quality of the fit indicates that variation of N_s within the range 0.6 to 1.0 does not greatly influence the value of the coefficient α.

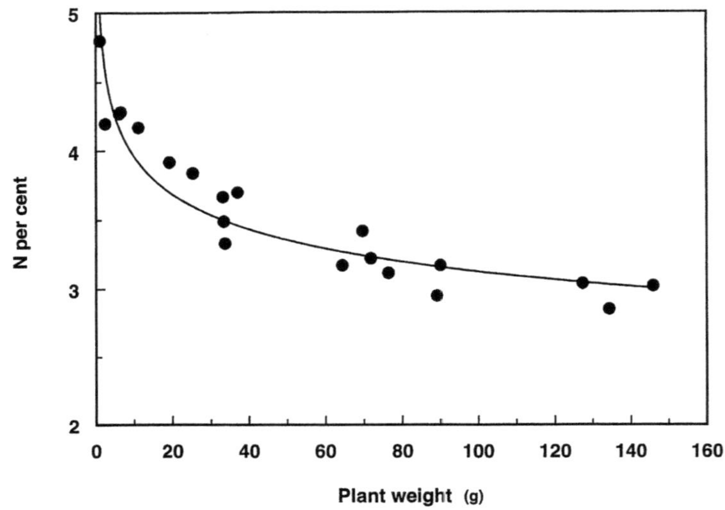

Fig. 1.8. Relationships between plant N% and plant mass (W) for isolated sweet sorghum plants. The two fitted equations: $N\% = 5.06(W)^{-0.106}$ and $N\% = 4.27(W)^{-0.128} + 0.8$ are comfounded and are represented by the same line

This adjusted value of α (0.872) is close to the value of 0.855 calculated directly from Eq. (20). It is important to note here that Eqs. (19) and (19′) do not allow estimation of coefficient N_m independently of c or W_i. This reflects the difficulty in precise definition of the metabolic and structural plant components, which remain only conceptual entities.

A first attempt to make these two conceptual plant components more concrete would be to consider that leaf tissues correspond more to the metabolic component, and stem supporting tissues more to structural material. Figure 1.9 shows the decrease in N% for the two morphological components of plant biomass, lamina fraction and pseudostem fraction, for isolated plants of sweet sorghum. The parallel decline in N% in these two organs suggests that they accumulate structural tissues in equal proportion. In lamina, as shown by Fig. 1.7, the midrib represents a greater proportion of mass as the lamina becomes larger. Thus, the decline in lamina N% is the consequence of the increased proportion of midrib tissues with low N% compared to the nonmidrib tissue of the lamina. As the leaf area per plant increases, the leaf N content expressed on a dry matter basis, that is N%, declines, while the leaf N content expressed on an leaf area basis remains more or less constant at about $1.6\,g\,N\,m^{-2}$.

The N% decline in pseudostem fraction is more difficult to analyze because of the heterogeneity of this morphological compartment in *Gramineae*. Pseudostems include mature sheath and stem tissues with a low N%, and meristematic tissues of growing leaves and growing internodes with a high N%. The proportion of mature tissues in pseudostem increases as the plant accumulates mature intern-

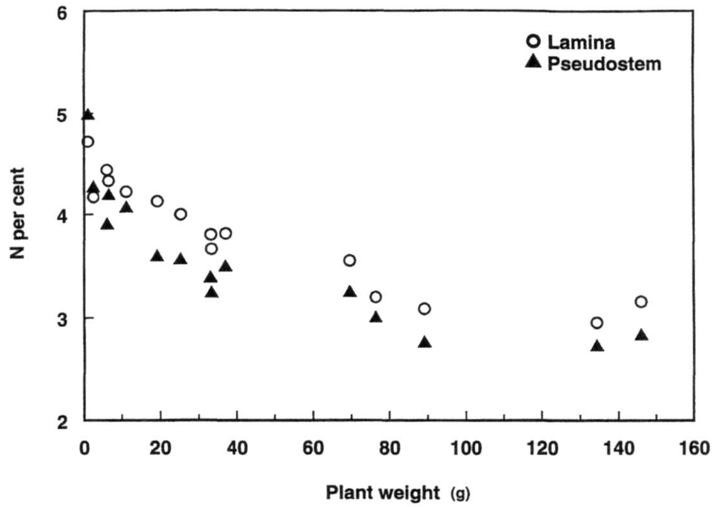

Fig. 1.9. Decline in N% with increasing plant mass for lamina and pseudostem fraction of isolated plants of sweet sorghum

odes, which leads to a rapid decrease in stem N% with increasing stem mass. Thus, the metabolic plant part could correspond to chloroplastic tissues plus expanding meristematic tissues with a high N concentration, while the structural component could correspond to mature supporting tissues with a low N concentration. In monocotyledons, these two types of tissues overlap in both the leaf and stem fraction, and therefore leaf weight ratio (LWR) does not appear to be a good estimator of the proportion of metabolic part in the plant. In dicotyledons such as *Medicago sativa*, the leaves are mainly composed of chloroplastic mesophyll tissues, and the supporting tissues are essentially included in stem internodes. For this reason, in dicotyledons, LWR appears to be a good indicator of the proportion of metabolic issues in the plant, and is therefore closely linked to plant N%.

The model and the data presented above give a theoretical basis for explaining the decrease in critical plant N% with increasing plant mass. The basic assumptions of this model are that the metabolic activity of the plant is determined by the proportion of plant biomass which is associated with light interception and conversion, and that the N demand for maximum plant growth corresponds to the minimum N necessary to achieve maximum leaf area expansion. However, as leaf area develops, plants have to invest a greater proportion of their carbon gain in structural tissues in order to maintain the extra leaf area in an adequate position for intercepting more light. The model developed in this section is necessary to understand how the presence of neighboring plants modifies the access of leaves to light, and hence the growth pattern and N demand of individual plants growing in a dense canopy.

1.3.3
Effects of Competition for Light on Plant N% Decline in Dense Crops

Comparison of the decline in plant N% with increasing plant mass between isolated plants and the average individual plant in a dense canopy allows the overall effect of competition between plants to be quantified. Figure 1.10 shows that for a mean mass per plant greater than around 10 g, the slope of the relationship between ln(plantN%) and ln(plant mass) for an individual plant declines more rapidly in a dense canopy than in an isolated plant. This point corresponds approximately to a crop biomass of 150 $g\,m^{-2}$. After this point the slope of the regression line for individual plants in a dense canopy rapidly reaches a value of –0.34, which corresponds to the general value found for coefficient b in Eq. (1), while the slope for isolated plants retains a constant value of –0.106.

It is clear that competition for light between individual plants inside a dense crop causes a large reduction in plant N%. In the data presented in Fig. 1.10, competition for soil N cannot be put forward as a cause of plant N% decline, because plant N% corresponded to the critical plant N concentration, and additional N supply did not result in extra growth. The leaf area index (LAI) of the crop at the point at which the regression line between ln(plant N%) and ln(plant mass) diverged from that of isolated plants was approximately 2.5, which corresponds to an interception of about 77% of incoming radiation.

A first hypothesis for explaining the more rapid decrease in plant N% in dense canopies can be found in the relationship between the distribution of leaf N and

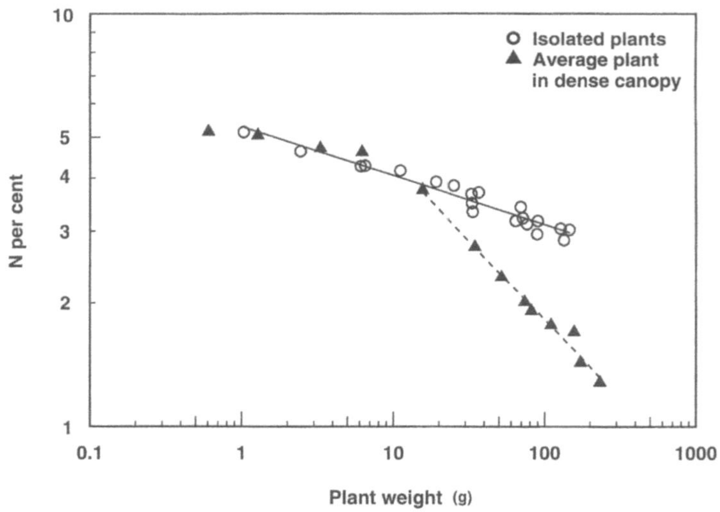

Fig. 1.10. Comparison of the relationship between plant N% and plant mass between isolated plants and the average plant in a dense canopy for sweet sorghum cv. Keller. The data for the average plant in the dense canopy has been obtain by dividing the crop mass per unit ground area by the plant density at each harvest date

irradiance within plant canopies. Several authors have shown that within dense canopies of different species, leaf nitrogen partitioning is nonuniform, and parallels the vertical light distribution (Field 1983; Charles-Edwards et al. 1987; Hirose and Werger 1987; Pons et al. 1993). When leaf N content is expressed on a leaf area basis, a close relationship can be found between the irradiance (I_z) at a given depth (z) within the canopy and the N content per unit leaf area (N_z) (Hirose et al. 1988; Lemaire et al. 1991), as shown in Fig. 1.11. The nonlinearity of such relationships suggests the existence of two modifying kinetics for N_z as a function of I_z: a gradual decrease with light extinction down to a value of I_z, corresponding to the light compensation point of daily net photosynthesis, and a rapid decrease beyond this point, when the carbon balance of any individual leaf would be negative and protein maintenance would no longer be possible, leading to leaf senescence, yellowing, and abscission.

The minimum values observed for N per unit of leaf area at the base of a lucerne canopy shown in Fig. 1.11 corresponds to approximately 30% of the leaf nitrogen content of the well-illuminated leaves at the top of the canopy. This represents a loss of about 70% of the leaf N as the leaf's relative position within the canopy changed during canopy regrowth. Measurement of N content per unit leaf area on yellow leaves collected immediately after leaf drop indicated that 80% of the initial leaf N content is recycled before leaf abscission (Lemaire et al. 1991). The fraction of leaf N remaining in dead leaf presumably corresponds to structural leaf N, whereas the fraction remobilized corresponds to the metabolic pool.

The remobilization of metabolic leaf N from shaded leaves to well-illuminated leaves corresponds to an optimization of canopy photosynthesis, as suggested by Mooney and Gulmon (1979). The relationship between leaf net photosynthesis and

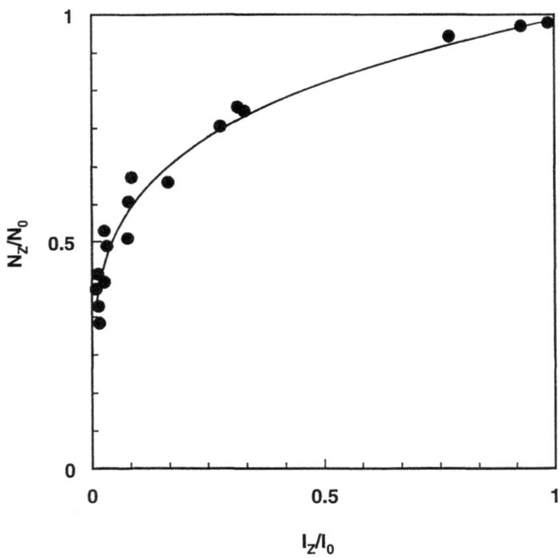

Fig. 1.11. Relationships between the relative irradiance at depth *z* within the canopy (Iz/Io) and the relative nitrogen content per unit leaf area (Nz/No) for a lucerne stand. Io and No represent the irradiance and leaf N content at the top of the canopy. (After Lemaire et al. 1991)

leaf N content shows that leaves of high N content saturate at higher irradiances and have higher light-saturated net photosynthesis rates than leaves with low N content, but at low irradiance there is little or no difference (Gulmon and Chu 1981; Hirose and Werger 1987; Sage and Pearcy 1987). Thus, the maximum daily net photosynthesis for a given quantity of N in canopy is achieved with a preferential allocation of N to the well-illuminated horizon of the canopy.

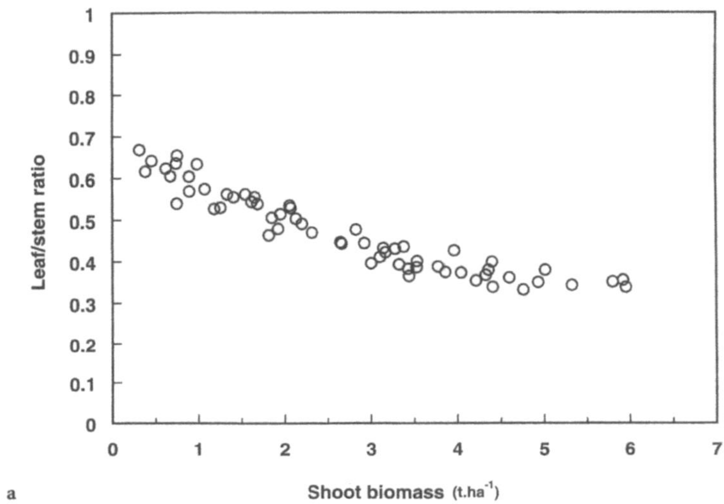

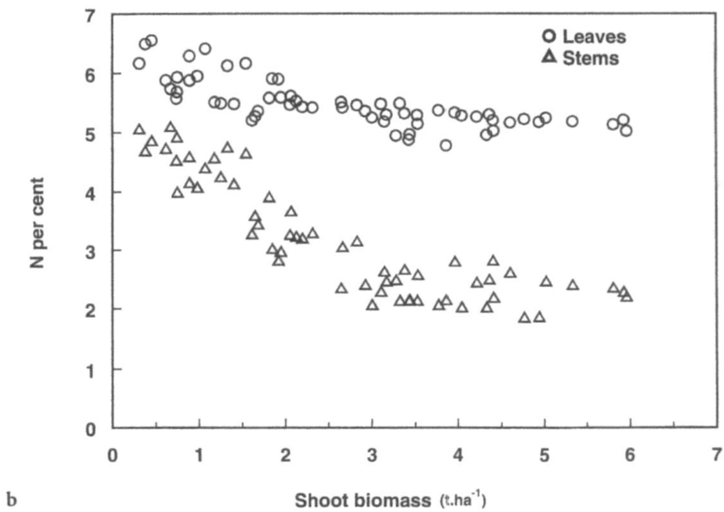

Fig. 1.12. **a** Decrease in leaf/stem ratio (L/S) of lucerne stands with increasing crop mass. **b** N% decline in leaves of stems of lucerne with increasing crop mass

This nonuniform leaf N distribution within plant canopies could provide an explanation for the lower plant N% of plants growing in dense canopies than isolated plants when the comparison is made at the same plant mass. Lemaire et al. (1992) confirmed that the mean leaf N content per unit leaf area (N_L $g\,m^{-2}$) within a lucerne crop declines as the leaf area index of the canopy (LAI) increases and as a greater proportion of leaves becomes shaded:

$$N_L = 2.36(LAI)^{-0.22} \qquad R^2 = 0.97.$$

However, such a decrease is accompanied by a decrease in specific leaf weight (SLW $g\,m^{-2}$):

$$SLW = 39.4(LAI)^{-0.15} \qquad R^2 = 0.95.$$

As a consequence, leaf N% expressed on a dry weight basis decreases only slightly with the increase in the LAI of the crop. Thus, as shown in Fig. 1.12, the decline in plant N% with increasing crop biomass in lucerne is mainly due to, (1) a decrease in N% in stem fraction, and (2) a decrease in the leaf: stem ratio, while the decline in mean leaf N% remains relatively limited. The major factor responsible for the decline in plant N% in a dense canopy is the increasing proportion of structural tissues.

Figure 1.13 allows a comparison of the dynamics of leaf and stem growth for sweet sorghum plants growing either isolated or in a dense canopy. It corresponds to data presented in Fig. 1.10. As the competition for light in a dense canopy increases, individual plants develop a greater proportion of stem tissues. The divergence with isolated plants starts at the same crop mass and crop LAI as that

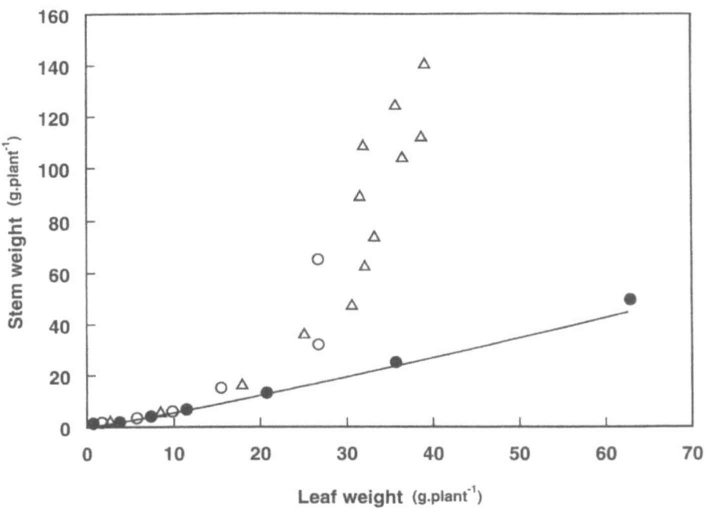

Fig. 1.13. Relationship between stem and leaf weight increase for isolated plants (● 1991) or average plants in a dense canopy (△ 1990; △ 1991) in sweet sorghum

shown for plant N% in Fig. 1.10. Figure 1.14 shows that the decline in plant N% in a dense canopy parallels the decline in leaf area ratio (LAR) of the plant, and a close relationship has been found between these two variables:

$$N\% = 285\,LAR + 0.21 \qquad R^2 = 0.989.$$

This relationship shows dependence of N% on LAR. It has been previously discussed by MacDonald (1989). The consequence of this linear relationship is that

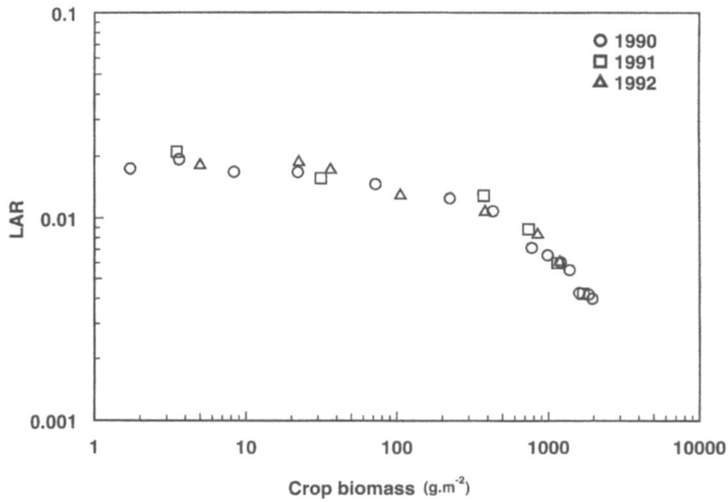

a

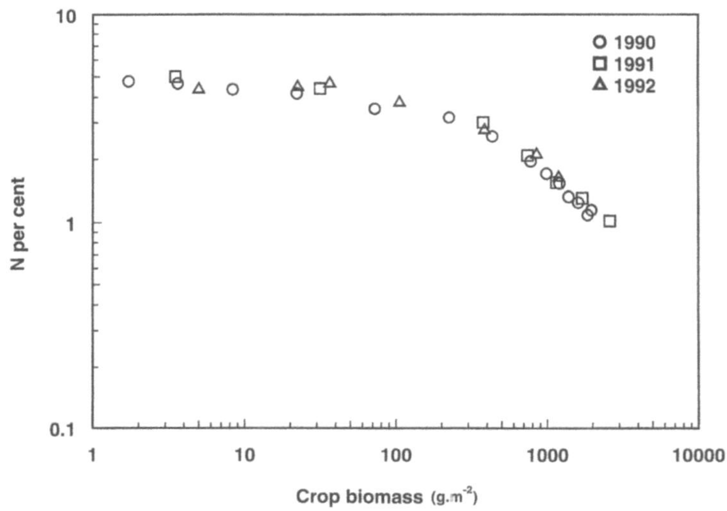

b

Fig. 1.14a–c. Parallel decreases in LAR (**a**) and plant N% (**b**) with increasing crop mass in sweet sorghum crops, and linear relationships between plant N% and LAR (**c**)

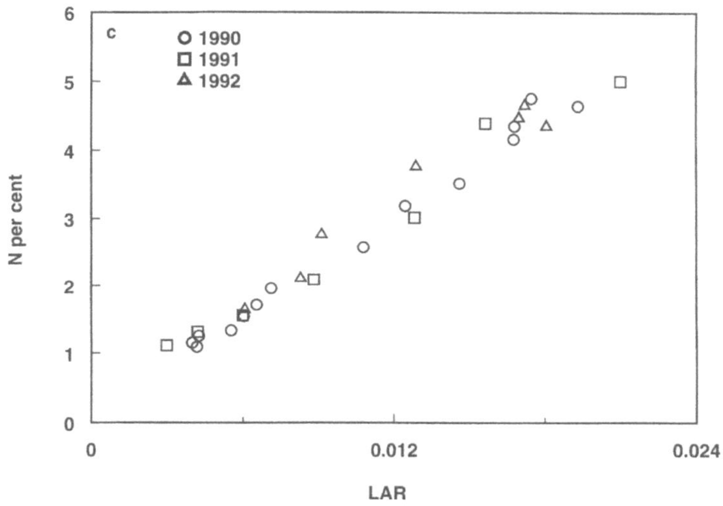

Fig. 1.14c

N accumulates in dense crops in proportion to increase in LAI. This idea of proportionality between the quantity of N accumulated and the green leaf area developed by a crop was previously proposed by Sylvester-Bradley et al. (1990) and Grindlay et al. (1993). The relationship implies that N in the structural tissues necessary to support leaf area, i.e., in stems and midribs, accumulates in proportion to leaf area and not to structural tissue mass. If we accept that the major part of the N in stem and midrib tissues is not directly linked to structural material per se, as cell wall component, but corresponds to N circulating in xylem and phloem tissues, we might propose that the quantity of N in nonleaf tissues is closely related to the leaf area itself as insuring exchanges of N compounds between leaves and roots and between shaded leaves and well-illuminated leaves. This idea is illustrated by Fig. 1.15, where it is shown that N accumulates in a lucerne crop at a constant rate of about 3 g N m^{-2} of leaf, but as the LAI increases, less and less of the extra N is used in the leaf component. These data correspond to those presented by Lemaire et al. (1991), which showed that the N content of individual stems of lucerne depended on their hierarchical position in the canopy, bigger stems bearing the most illuminated leaves having a greater N content. As the canopy develops, a greater proportion of the N is incorporated into stem fraction for supporting a canopy of a more or less constant leaf area, as new leaves appear at the top of the canopy and shaded leaves senesce.

The rapid decrease in leaf: stem ratio of individual plants in a dense canopy can be interpreted as a direct photomorphogenetic response to shading. This corresponds to a general increase in internode length mediated by the phytochrome equilibrium according to the decrease in red/far red ratio as the canopy becomes denser (Smith 1982). Thus, as shown in Fig. 1.13, the average individual

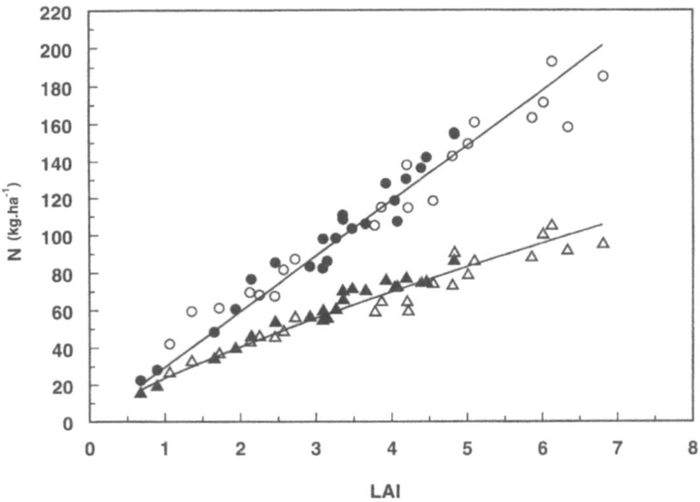

Fig. 1.15. Relationship between N accumulation in shoot (●, △) and leaves (▲, △) and LAI of lucerne during first growth in spring (*open symbols*) or regrowth in summer (*closed symbols*). Fitted equation for shoot: N = 29.5 (LAI) + 0.05, and for leaves: N = 21.8 $(LAI)^{0.85}$

plant in a dense canopy needs to develop a larger stem mass in order to position its leaves in the more illuminated layers of the canopy.

Figure 1.16 allows an analysis of the variation of N% among individual plants of different sizes within a plant canopy. At a given date of harvest, the largest plants in a dense canopy have a greater N% than smaller ones, despite their larger plant mass, which is in contradiction with data obtained on isolated plants, where larger individuals have a lower N% than smaller ones. Thus, due to the large variability in shoot height between individual plants in dense crops, there is a large variability in plant N% between individuals as a consequence of their hierarchical position and relative irradiance within the canopy. As demonstrated by several authors (Macduff and Wild 1988; Gastal and Saugier 1989; Rufty et al. 1989), the capacity of individual plants to absorb NO_3^- or to fix N_2 depends on the daily C supply. Thus, the less dominant plants in dense canopy, i.e., the smallest ones, which receive only limited irradiance, cannot absorb as much N as dominant plants, and their N% is greatly reduced in comparison to plants of similar size growing in the absence of shading. For dominant plants, i.e., the larger ones, the difference in N% compared to isolated plants of the same size is less as a consequence of their lower degree of shading. The dotted line in Fig. 1.16 represents the regression line of plant N% vs. plant weight for the average plant in a dense canopy. This line corresponds to the regression shown on Fig. 1.10 in log-log scale. As shown previously in Fig. 1.10, the intersection of the two lines corresponding to an isolated plant and to the average plant in a dense canopy occurs at a mean plant weight of 10 g, or a crop biomass of 1.50 t ha^{-1}. Before this stage, the decrease in N% of individual plants within the crop

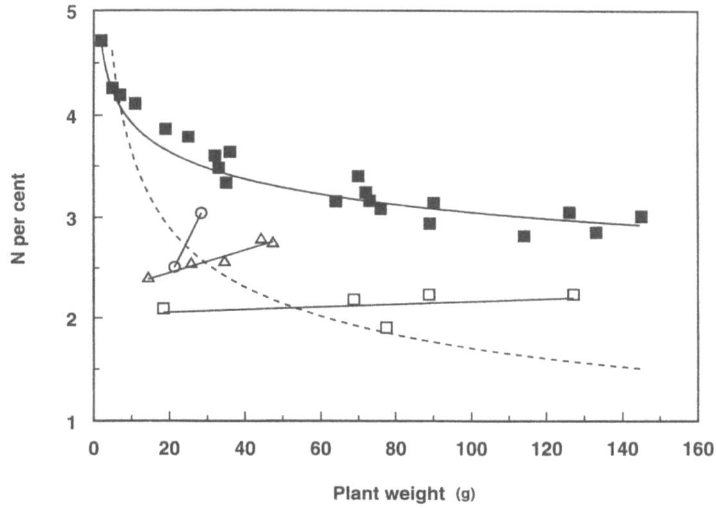

Fig. 1.16. Relationships between plant N% and plant mass (W) for an isolated plant of sweet sorghum (*closed symbols*) and for individual plants in a dense crop harvested at different dates (*open symbols*). The *continuous line* represents the equation N% = $5.06(W)^{-0.106}$ for isolated plants as in Fig. 1.9. The *dotted line* represents the equation N% = $7.79(W)^{-0.34}$ for the average plant in a dense canopy and corresponds to the dotted line of Fig. 1.10

can be approximated to that of isolated plants, and the crop N% decreases with increasing crop mass as follows:

$$N\% = 3.6(W)^{-0.106} \qquad \text{for } 0 < W < 1.5 \text{ t ha}^{-1}.$$

For a crop biomass higher than 1.5 t ha^{-1}, the decline in crop N% is more rapid and follows the average plant N%:

$$N\% = 3.6(W)^{-0.34} \qquad \text{for } W > 1.5 \text{ t ha}^{-1}.$$

As the total crop weight per unit ground area increases, the smallest plants, which contain less nitrogen because they are shaded, will be unable to grow and will probably die, leading to self-thinning. The theory developed above is in good agreement with the "skin and core" hypothesis proposed by Hardwick (1987) which postulates that metabolic N is associated with the illuminated area of canopy (skin), while plant mass is associated with the total volume of plant canopy (core), leading to an allometric coefficient of about 2/3 between the amount of N accumulated in a crop and the crop biomass.

The empirical model of Eq. (2), which allows us to relate crop N and crop biomass by an allometric relationship with a coefficient of around 2/3 whatever the species, appears to be a "law" imposed by geometrical constraints governing light extinction in dense canopies. Until now, we have had no idea if the plant N% vs. plant mass relationship for isolated plants could vary or not between species of the same metabolic pathway, C_3 or C_4. However, according to differences in plant morphology and architecture, we can imagine that the coefficient α of Eq. (16) could vary, reflecting intrinsic differences in leaf area ratio (LAR) between species. We have to note here that intrinsic differences in LAR, i.e., difference in LAR for a similar plant weight (W), must imply differences in RGR for isolated plants. Several authors have shown that genotypical differences in RGR exist for isolated plants growing in nonlimiting conditions (Konings 1989; Poorter 1989). Thus, the hypothesis that the slope of the regression line between ln(plant N%) and ln(plant mass) for isolated plants could be species-dependent seems reasonable. The main conclusion we can deduce from models and the data presented above is that whatever the genotypical differences for isolated plants, the regression line between ln(plant N%) and ln(plant mass) converges assymptotically to a value not very different from $-1/3$ for all species as plants grow in dense canopies. Thus the coefficient b of Eq. (1) would be species-independent and should be considered as a dense canopy property which reflects the universal exponential light extinction within canopy deduced from Beer's law. The differences observed in light extinction between species with erect or horizontal leaves are probably counteracted by the greater structural material associated with an erect canopy. Thus, the relative convergence of coefficient b for a large range of species could be interpreted as the result of optimization of both N distribution and structural material for light interception and light use by individual plants in dense canopies.

The decrease in plant N% seems to be driven by increase in plant size. Isolated plants increase their size mainly in two dimensions for the optimization of light capture. The allometry between leaf area and plant weight determines a parallel decline in LAR, plant N%, and RGR as plant size (or plant weight) expands. In a dense canopy, the competition for light between individual plants imposes on the plants a substantial increase in size in the third (vertical) dimension to maintain their leaf area in the well-illuminated layer of the canopy. This change in the pattern of expansion of the plant means that the decrease in both LAR and N% becomes more rapid as the canopy develops. If, to survive in a dense canopy, a plant has to maintain its leaf area at a threshold value of irradiance corresponding to its compensation point, dominated plants which do not succeed in increasing their height will die and the canopy will be subjected to a self-thinning process, leading to the so-called $-3/2$ power rule relating the plant density and biomass per surviving plant (Yoda et al. 1963). The skin and core hypothesis formulated by Hardwick (1987) assumes isometry between active, metabolic surface area and biomass for the surviving plants. Although White (1981) failed to find any single linear dimension that scaled isometrically with density and biomass in thinning stands, Sackville-Hamilton et al. (1995) showed that because (population mean biomass)$^{2/3}$ does not equal the mean for (individual biomass)$^{2/3}$, and because it is the smallest plants that die, the $-3/2$ rule relates to mean isometry of a changing population of surviving plants. Thus, in a dense canopy, size variation between plants is not isometric, but the average plant population tends to grow isometrically. Thus, the LAI of the stand tends to grow in proportion to (biomass)$^{2/3}$. The minimum N for maximum growth rate then evolves in a dense crop in proportion to (crop mass)$^{2/3}$, and critical plant N% varies proportionally to (crop mass)$^{-1/3}$.

This value of −1/3 is only a theoretical assumption for the coefficient b of Eq. (1). Fitted values obtained for different crops often give values close to −0.33. Values obtained include −0.32 by Lemaire and Salette (1984a) on *Festuca arundinacea* Schreb. swards; −0.29 to −0.36 by Lemaire et al. (1985) on *Medicago sativa* L.; −0.37 by Plenet (1995) on maize. A lower value of −0.44 has been found by Justes et al. (1994) for wheat. Greenwood et al. (1990) showed a common value of −0.5 for a large range of C_3 and C_4 crops, but because of the strong correlation between the coefficient a and b of Eq. (1), a value of −0.34 for b fits the data well if we exclude the data points corresponding to accumulation of storage material (grain or tuber filling).

Thus, the empirical equations (1) and (2) seem satisfactory for determining the critical plant N% and minimum N requirement for any crop at any growth stage. These empirical relationships are clearly supported by theoretical models relating plant N demand with the shape of plant size increase as a result of competition for light between plants inside dense canopies. The theory presented above accounts for the small variation of the coefficients of these relationships between species. The consistency of this 2/3 relationship with the −3/2 power rule gives it the same status of ecological law. Such a relationship can therefore be used as a reference for the determination of the N nutrition status of crops.

1.4 Determination of an Index of N Nutrition

1.4.1 N Nutrition Index of Crops

Models and theory presented above allow the validation of the Eq. (1) as an acceptable simplification of Eqs. (14) or (17) to describe the decrease in critical plant N% with increasing crop biomass. From the point of view of diagnosis, we have to choose the simplest equation, and we have shown in preceding sections that the introduction of the value of N_s as an asymptote does not significantly improve the quality of the fit. Furthermore, because of the unknown value of N_s for each species, it appears reasonable to omit it. When a crop can be considered as a closed canopy, i.e., for crop biomass greater than 1.0–1.5 t ha^{-1}, Eq. (1) can be used. Before this point, in the earlier stages of crop growth, the plants can be considered as isolated, and plant N% declines slowly, as shown in Fig. 1.10. During this early growth period, Eq. (19) has to be used to estimate the decline in plant N%; but we have to note here that for the very young stages of growth, the critical plant N% is not defined well by the model because initial plant N% would be equal to N_m when total plant mass is composed only of the metabolic component. Therefore, it is necessary to introduce initial conditions in the model to account for the plant mass W_i at the beginning of the autotrophic growth. In the absence of adequate data to allow exact determination of the end of the heterotrophic growth of seedlings, however, we can neglect this early phase of crop growth and the optimum plant

N% during this period can be considered as constant and equal to the value of coefficient a in Eq. (1), i.e. the plant N% for a crop mass of 1.0 t ha^{-1}.

Thus, critical plant N% can be estimated at any stage of growth by the two following equations:

$$N\% = a \qquad \text{for } W < 1.0 \text{ t ha}^{-1}.$$

$$N\% = a(W)^{-b} \qquad \text{for } W > 1.0 \text{ t ha}^{-1}.$$

Such an evaluation of critical plant N% is valid only for the period of vegetative growth of the crop. When crop growth involves not only growth of leaf and stem tissues, but also accumulation of storage materials associated with grain or tuber filling, the relation between plant N% and crop biomass depends mainly on the biochemical nature of the storage materials. For cereal crops or potatoes, which accumulate mainly starch in grains or tubers, plant N% would decline more rapidly with increasing crop mass as grain or tuber filling occurred, while for grain legume crops, which accumulate large amounts of proteins in the grains, the decline in plant N% would be comparatively slower as grains develop. Thus it is important to stress here the necessity to restrict the use of Eq. (1) for determining critical plant N% exclusively to periods of vegetative growth, and to be cautious in using the above equation during periods of grain filling.

One question which arises with the use of Eq. (1) is the analytical determination of plant N%. According to the theory developed above, the reduced N concentration would certainly be a better way to determine plant N% than the total N concentration. Kjeldahl analysis remains the most commonly used method to determine plant N%, but this method can include a variable fraction of nitrate N depending on the digestion procedure (Guiraud and Fardeau 1977). Nevertheless, Justes et al. (1994) showed that the critical curves for total N and for reduced N for wheat were very close for crop biomass >1.5 t ha^{-1}. At a low crop biomass, plants can contain a significant amount of N in nitrate form, and it is recommended that critical plant N% be determined by the analysis of reduced N by substraction of nitrate N from total N (Dumas' method). Determination of the reduced N in this way should be preferable for crops which are considered to accumulate large quantities of nitrate, such as rape, or some vegetable crops, such as spinach or lettuce.

Assuming that Eq. (1) describes the decline in critical plant N% with increasing crop mass, a nitrogen nutrition index (NNI) can be calculated at any time of crop growth as the ratio between the actual plant N% (N_{act}) and the critical plant N% (N_c) that the crop should have at that biomass as proposed by Lemaire et al. (1989) and illustrated in Fig. 1.17:

$$NNI = \frac{N_{act}}{N_c} \tag{21}$$

A value of NNI greater than or equal to 1 indicates that the crop was in a situation of non-limiting N supply at the date of harvest, i.e., N supply does not limit the growth rate of the crop at this time. Such an index allows a diagnosis of

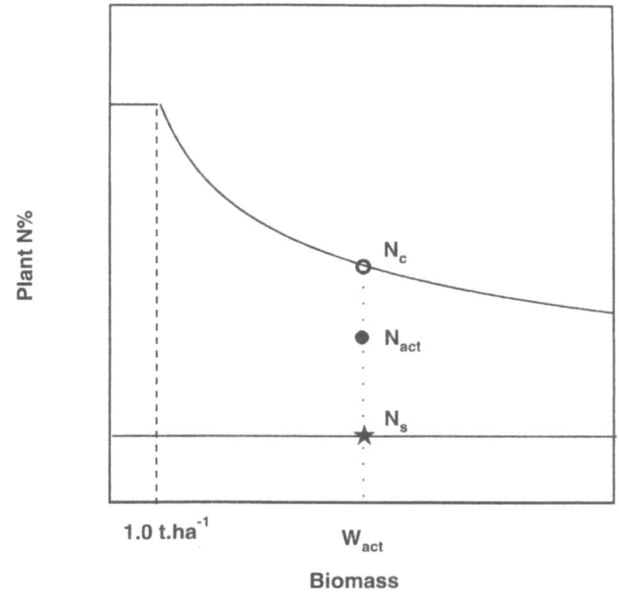

Fig. 1.17. Diagram for the determination of the nitrogen nutrition index: $NNI = N_{act}/N_c$, $NNI' = (N_{act} - N_s)/(N_c - N_s)$

the existence, or not, of a nitrogen deficiency in the crop at the time of observation, and an evaluation of the intensity of this deficiency.

The NNI is similar to the index proposed by Agren (1985), but the difference is that this author considered N_c as constant during the early, exponential period of growth. Angus and Moncur (1985) proposed a similar index but, using maximum plant N% as the reference. Greenwood et al. (1986) also suggested a calculation of an N index using optimum and minimum N% values as references; optimum N% was related to crop mass by an empirical negative exponential equation, while minimum N% was assumed to be constant and represented the minimum plant N% for plant growth. These authors proposed the calculation of crop N status as the ratio between (actual N% minus minimum N%) and (optimum N% minus minimum N%). It is interesting to note that if we introduce an assymptotic value of N_s for plant N%, as shown in Eq. (19), we should be able to calculate a similar nitrogen nutrition index:

$$NNI' = \frac{N_{act} - N_s}{N_c - N_s}. \tag{22}$$

NNI′ can vary from 0 in very deficient conditions, to values greater than 1 for luxury consumption of N. The NNI can vary only from a minimum value equal to N_s/N_c, to values greater than 1 and is therefore less discriminating for N-limiting conditions. Because N_s is theoretically a constant, the minimum value of NNI increases as N_c decreases, i.e., as crop mass becomes bigger. Therefore, similar values of NNI do not necessary represent exactly the same degree of N deficiency, depending on the crop biomass at which NNI is calculated: but we have no indication to date about the exact value of N_s for different species, and, as mentioned before, a large range of values between 0.5 and 1 could be acceptable. Thus, in the absence of any precision on the value of N_s, we suggest the use of NNI and not NNI′ as the index of N nutrition status of crops.

The procedure for estimation of the value of NNI is very simple:

1. Estimation of actual crop mass (W_{act}) by taking crop samples.
2. Measurement of actual plant N% (N_{act}) after grinding a sub-sample of harvested plant material. As discussed previously, N determination can be carried out with the Kjeldahl or the Dumas procedures. In the latter case, it will be preferable to determine nitrate-N content in addition, to express the result as content of reduced N for situations where nitrate accumulation in plants might occur.
3. Calculation of the critical plant N% (N_c) corresponding to the observed crop mass (W_{act}), using Eq. (1) and acceptable values of coefficients a and b for the species considered when fitted equations have been proposed in litterature. Since it has been developed from the theoretical basis presented above, the diagnosis of crop N nutrition with the index NNI should be of general validity, at least for species in which coefficients a and b of Eq. (1) have been determined with a sufficient statistical precision.

1.4.2
Relationship Between the N Nutrition Index and Crop Biomass Production

The definition of the critical plant N% given in Section 1.2.2 implies that, as the actual plant N% decrease below the critical value, the crop growth rate falls below its maximum value. In consequence, the regression line between critical plant N% and crop biomass as approximated by Eq. (1) represents the plant N%-crop mass trajectory for maximum biomass accumulation. A crop whose trajectory is always above or just on this line should accumulate the maximum biomass allowed by other environmental conditions (temperature, radiation, water, and mineral supply) during the period considered. As the trajectory of plant N% trajectory decreases below the critical line, the growth rate of the crop decreases proportionally to the distance between the actual and critical plant N%, that is estimated by the value of the nitrogen nutrition index. Thus we can assume a linear relationship between actual vs. maximum crop growth rate ratio and instantaneous values of NNI of the crop for NNI < 1:

$$\frac{\left(\frac{dW}{dt}\right)_{act}}{\left(\frac{dW}{dt}\right)_{max}} = a + b\text{NNI}. \tag{23}$$

By definition, $(dW/dt)_{act} = (dW/dt)_{max}$ when NNI is greater than or equal to 1. In field conditions, instantaneous crop growth rates cannot be estimated easily. Nevertheless, some components of the growth can be estimated at daily intervals and can be related directly to the instantaneous value of NNI. Thus, Gastal et al. (1992) showed a linear relationship between the daily leaf elongation rate and the NNI of tall fescue swards grown with different N application rates.

The actual biomass (W_{act}) accumulated by a crop at a date of harvest t can be considered as the integral of the instantaneous crop growth rate during the period t-t_0. Therefore, the ratio W_{act}/W_{max}, which represents the ratio between the biomass accumulated by a crop receiving a limiting N supply and the biomass accumulated by a crop under optimum N supply, must be related to the integrated values of NNI less than or equal to 1 during the period $t - t_0$. Calculation of the integrated value of the nitrogen nutrition index during a period of growth implies a knowledge of the time course of the NNI during this period. As shown in Fig. 1.18, which is for tall fescue swards receiving different N application rates, the instantaneous NNI decreases during the regrowth period as the available soil N is progressively absorbed.

Figure 1.19 shows the relationship between the relative accumulated biomass (W_{act}/W_{max}) and the nitrogen nutrition index for swards of tall fescue grown with different N application rates. When W_{act}/W_{max} is plotted against instantaneous NNI determined at harvest as defined in Eq. (21), the relationship obtained is obviously nonlinear (Fig. 1.19a). When W_{act}/W_{max} is plotted against the integrated value of NNI during the period of regrowth, the relationship appears to be linear (Fig. 1.19b). The difference between the two methods of calculating NNI reflects the fact that an identical NNI at the date of harvest can result from different NNI trajectories during the period of regrowth, leading to differences in the integrated NNI value. Thus, knowledge of the time course of the NNI during the period of growth

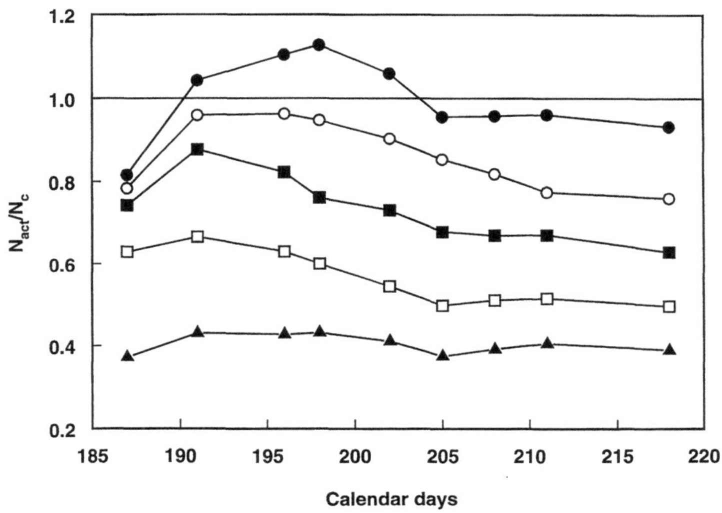

Fig. 1.18. Evolution of the NNI of tall fescue swards during a summer regrowth with irrigation and for different levels of N fertilizer: 160 (●); 120 (○); 80 (■); 40 (□); 0 (▲) kg ha^{-1}

considered is as important as knowledge of the final value of NNI at the date of harvest. As illustrated by the example of the Fig. 1.18, the use of the instantaneous NNI instead of the integrated NNI could lead to an exaggeration of the severity of the overall N deficiency experienced by the crop during the period of regrowth, in particular for the less deficient treatment, where the NNI falls under a value of 1 only at the end of the period. This distortion in estimating the overall N deficiency during the period of regrowth leads to the concavity of the response shown in Fig. 1.19a in comparison with the linear response shown in Fig. 1.19b. The positive

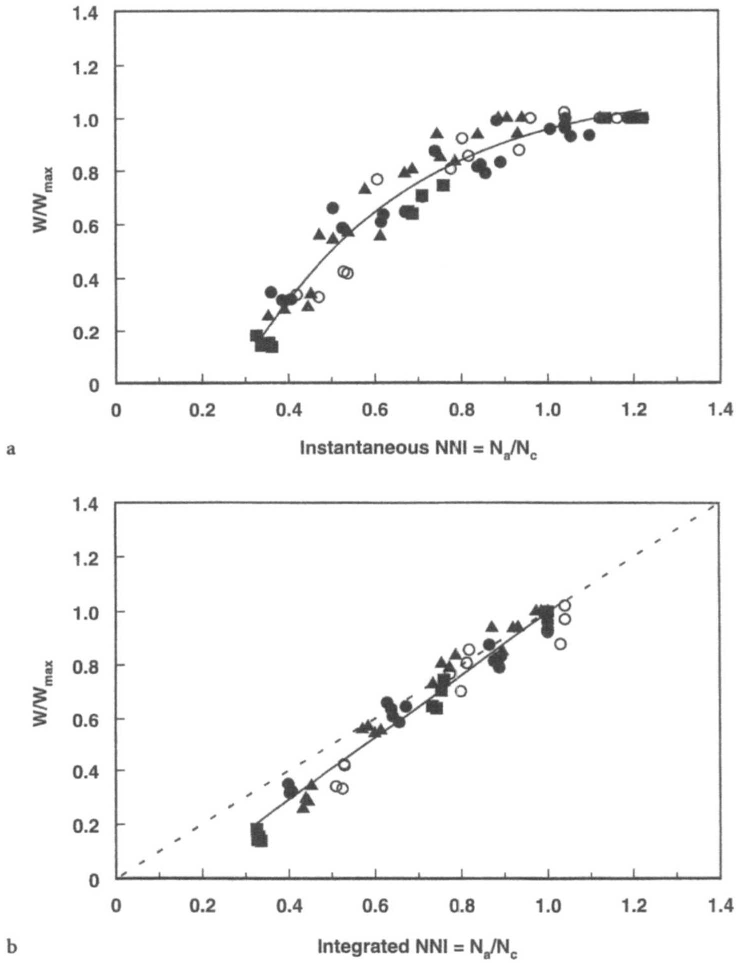

Fig. 1.19a–c. Relationships between the relative biomass accumulation (W/W_{max}) of tall fescue swards growing in spring 1988 (●) and 1989 (○) and in summer 1987 (▲) and 1988 (■) with different levels of N applications and the nitrogen nutrition index of the swards calculated as **a** instantaneous NNI = N_a/N_c; **b** integrated NNI = N_a/N_c and **c** integrated NNI′ = $(N_a - N_s)/(N_c - N_s)$ with $N_s = 0.8$

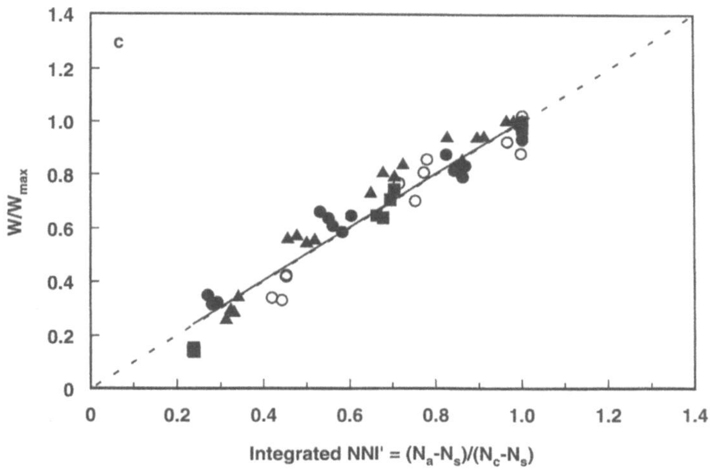

Fig. 1.19c

intercept on the *x* axis indicates that a minimum NNI is necessary for growth to occur; but, as indicated above, this minimum value can vary according to the corresponding crop mass. Introduction of a value of $N_s = 0.8$ in the calculation of the NNI′ as proposed by Eq. (22), allows the regression line to pass through origin and to be statistically confounded with $y = x$ (Fig. 1.19c).

Two mistakes have to be avoided in the use the N nutrition index. The first is to use only instantaneous values of NNI to explain observed differences in crop mass. In the examples shown in Fig. 1.19, the error due to using an instantaneous as opposed to an integrated value of NNI could be considered as not very large because, as shown in Fig. 1.18, the change in NNI during the period of regrowth for each N fertilizer treatment is relatively small and regular, so a good correlation exists between instantaneous and integrated NNI values; but in conditions of very rapid changes in N supply, correlation could be low. Thus, an N fertilizer application after a period of N deficiency can rapidly restore the instantaneous NNI to a value of 1, but the integrated NNI would remain less than 1, and the biomass accumulated by the crop would be reduced in comparison with a crop maintained under a nonlimiting N supply during all the period.

The second mistake is to use an average value of NNI instead of the integrated value of $I_N < 1$. This would lead to an underestimation of the effect of a period of N deficiency when it followed a period of luxury consumption, as is frequently the case for crops receiving an N application at the beginning of the growth period. By definition, crops having a greater plant N% than the minimum required for the maximum growth rate do not have any advantage in terms of growth capacity. The only advantage for a crop of having luxury consumption of N is to delay the onset of N deficiency when soil N shortage occurs during later growth.

Thus, in situations where soil N supply varies rapidly during growth, frequent estimations of instantaneous values of NNI allow an unbiased integrated value of

$I_N < 1$ to be estimated, which could be related to relative crop biomass. For annual crops receiving N applications in the early phase of growth period, a large decrease in NNI is often observed during growth, and caution has to be taken in the sequencing of diagnosis during the growth period. Some crops, such as perennial grasses, show a relatively constant NNI during the regrowth period for a given N application rate, as shown in Fig. 1.18 for tall fescue swards. This phenomenon is due to the smoothing effect of both root N reserves and rhizospheric microbial activity on N availability for the plants in such an ecosystem. In this case, only a few determinations of instantaneous NNI value are necessary for a good estimate of the integrated NNI value.

1.5 Conclusion

The theory presented above allows the generalization of the use of the empirical relationship between critical plant N% and crop mass [Eq. (1)] as an indicator of the crop N status. The decline in plant N% with plant mass appears to be a universal process which results from the ontogenic decline of the proportion of metabolic vs. structural tissue as the size of the plant increases. The greater plant N% of C_3 compared to C_4 crops at a similar crop mass reflects both a higher N concentration in the metabolic tissues of the former, and a greater proportion of metabolic as opposed to structural components. Competition for light between individual plants in dense canopies leads to a rapid change in plant morphology, and provokes an acceleration of the decrease in LAR, which parallels the decline in plant N%. The data obtained seem to indicate that the N accumulation rate in a dense crop is proportional to the rate of increase in LAI, at least during the vegetative growth period; but because for many crops LAI tends to increase isometrically in relation to crop mass, i.e., in proportion to (crop mass)$^{2/3}$, the power coefficient of the relationship between plant N% and crop mass appears to be close to −1/3.

The constancy of the coefficient b over a large range of species must be confirmed by more precise determinations. Ontogenic changes in plant morphology could cause some variations in the allometric coefficient between crop LAI and crop mass, and therefore could lead to specific variations of the coefficient b from the theoretical value of −1/3. It is clear that the theory developed here implies a certain stability in the shape of the organs developed by plants. All the data presented concern vegetative crop development, i.e., plants which produce only leaves and stems. As dry matter accumulates in other types of organs of different shape and different density, the theory based on stable surface-volume allometry and a constant mass/volume ratio cannot be used. In consequence, we think that, despite the theory which supports a constant value of b = 1/3, it seems preferable to determine empirically the value of the coefficients a and b of Eq. (1) for each kind of crop in order to obtain a reference curve for the variation of critical plant N% to calculate the nitrogen nutrition index.

References

Agren GI (1985) Theory for growth of plants derived from the nitrogen productivity concept. Physiol Plant 64:17–28

Allirand JM, Gosse G, Lemaire G (1992) Influence of temperature on lucerne dry matter and nitrogen distribution. In: Scaife A (ed) Proc 2nd Congr of European Society of Agronomy, Warwick, pp 24–25

Angus JF, Moncur MW (1985) Models of growth and development of wheat in relation to plant nitrogen. Aust J Agric Res 36:537–544

Brown RH (1978) A difference in N use efficiency in C_3 and C_4 plants and its implications in adaptation and evolution. Crop Sci 18:93–98

Brown RH (1985) Growth of C_3 and C_4 grasses under low N levels. Crop Sci 25:954–957

Burns IG (1992) Influence of plant nutrient concentration on growth rate: use of a nutrient interruption technique to determine critical concentrations of N, P and K in young plants. Plant Soil 142:221–233

Caloin M, Yu O (1982) An extension of the logistic model of plant growth. Ann Bot 49:599–607

Caloin M, Yu O (1984) Analysis of the time course change in nitrogen content of *Dactylis glomerata* L. using a model of plant growth. Ann Bot 54:69–76

Charles-Edwards DA (1982) Physiological determinants of grop growth. Academic Press, Sydney, 161 pp

Charles-Edwards DA, Stutzel H, Ferraris R, Beech DF (1987) An analysis of spatial variation in the nitrogen content of leaves from different horizons within a canopy. Ann Bot 60:421–426

Ericksson T (1981) Effects of varied nitrogen stress on growth and nutrition in three *Salix* clones. Physiol Plant 51:423–429

Field C (1983) Allocating leaf nitrogen for the maximization of carbon gain: leaf age control on the allocation program. Oecologia 56:341–347

Field C, Mooney HA (1986) The photosynthesis-nitrogen relationship in wild plants. In: Givnish TJ (ed) On the economy of plant form and function. Cambridge University Press, Cambridge, pp 25–55

Gastal F, Saugier B (1989) Relationships between N uptake and C assimilation in whole plants of tall fescue. Plant Cell Environ 12:407–418

Gastal F, Belanger G, Lemaire G (1992) A model of the leaf extension rate of tall fescue in response to nitrogen and temperature. Ann Bot 70:437–442

Gosse G, Varlet-Grancher C, Bonhomme R, Chartier M, Allirand JM, Lemaire G (1986) Production maximale de matière sèche et rayonnement solaire intercepté par un couvert végétal. Agronomie 6:47–56

Greenwood DJ, Neeteson JJ, Draycott A (1986) Quantitative relationships for the dependence of growth rate of arable crops to their nitrogen content, dry weight and aerial environment. Plant Soil 91:281–301

Greenwood DJ, Lemaire G, Gosse G, Cruz P, Draycott A, Neeteson JJ (1990) Decline in percentage N of C_3 and C_4 crops with increasing plant mass. Ann Bot 66:425–436

Greenwood DJ, Gastal F, Lemaire G, Draycott A, Millard P, Neeteson JJ (1991) Growth rate and %N of field grown crops: theory and experiments. Ann Bot 67:181–190

Grindlay DJC, Silvester-Bradley R, Scott RK (1993) Nitrogen uptake of young vegetative plants in relation to green area. J Sci Food Agric 63:116

Guiraud G, Fardeau JC (1977) Dosage par la méthode Kjeldahl des nitrates contenus dans les sols et les végétaux. Ann Agron 28:329–333

Gulmon SL, Chu CC (1981) The effects of light and nitrogen on photosynthesis, leaf characteristics, and dry matter allocation in the chaparral shrub, *Diplacus auranticus*. Oecologia (Berl) 49:207–212

Hanway JJ (1962) Corn growth and composition in relation to soil fertility. III. Percentages of N, P and K in different plant parts in relation to stage of growth. Agron J 54:222–230

Hardwick RC (1987) The nitrogen content of plants and the self-thinning rule of plant ecology: a test of the core-skin hypothesis. Ann Bot 60:439–446

Hirose T, Werger MJA (1987) Maximising daily canopy photosynthesis with respect to the leaf nitrogen allocation pattern in a the canopy. Oecologia 72:520–526

Hirose T, Werger MJA, Pons TL, van Rheenen WA (1988) Canopy structure and leaf nitrogen distribution in stand of *Lysimachia vulgaris* L. as influenced by stand density. Oecologia 77:145–150

Ingestad T (1979) Nitrogen stress in birch seedlings. II. N, K, P, Ca and Mg nutrition. Physiol Plant 45:149–157

Ingestad T, Lund AB (1979) Nitrogen stress in birch seedlings. I. Growth technique and growth. Physiol Plant 45:454–466

Justes E, Mary B, Meynard JM, Machet JM, Thelier-Huché L (1994) Determination of a critical nitrogen dilution curve for winter wheat crops. Ann Bot 74:397–407

Konings H (1989) Physiological and morphological differences between plants with a high NAR or a high LAR as related to environmental conditions. In: Lambers H, Cambridge ML, Konings H, Pons TL (eds) Causes and consequences of variation in growth rate and productivity of higher plants. SPB Academic Publishing, The Hague, pp 101–123

Lemaire G, Chartier M (1992) Relationships between growth dynamics and nitrogen uptake for individual sorghum plants growing at different plant densities. Proc 2nd Congr of European Society of Agronomy, Warwick University, pp 98–99

Lemaire G, Salette J (1984a) Relation entre dynamique de croissance et dynamique de prélèvement d'azote pour un peuplement de graminées fourragères. I. Etude de l'effet du milieu. Agronomie 4:423–430

Lemaire G, Salette J (1984b) Relation entre dynamique de croissance et dynamique de prélèvement d'azote pour un peuplement de graminées fourragères. II. Etude de la variabilité entre génotypes. Agronomie 4:431–436

Lemaire G, Gruz P, Gosse G, Chartier M (1985) Etude des relations entre la dynamique de prélèvement d'azote et la dynamique de croissance en matière sèche d'un peuplement de luzerne (*Medicago sativa* L.). Agronomie 5:685–692

Lemaire G, Gastal F, Salette J (1989) Analysis of the effect of N nutrition on dry matter yield of a sward by reference to potential yield and optimum N content. Proc XVI Int Grassland Congr, Nice, pp 179–180

Lemaire G, Onillon B, Gosse G, Chartier M, Allirand JM (1991) Nitrogen distribution within a lucerne canopy during regrowth: relation with light distribution. Ann Bot 68:483–488

Lemaire G, Khaity M, Onillon B, Allirand JM, Chartier M, Gosse G (1992) Dynamics of accumulation and partitioning of N in leaves, stems and roots of lucerne (*Medicago sativa* L.) in a dense canopy. Ann Bot 70:429–435

MacDonald AJS (1989) Phenotypic variation in growth rate as affected by N-supply: its effect on net assimilation rate (NAR), leaf weight ratio (LWR) and specific leaf area (SLA). In: Lambers H (ed) Causes and consequences of variation in growth rate and productivity of higher plants. SPB Academic Publishing, The Hague, pp 37–44

Macduff JH, Wild A (1988) Changes in NO_3^- and K^+ uptake by four species in flowing solution culture in response to increased irradiance. Plant Physiol 74:251–256

Monteith JL (1977) Climate and the efficiency of crop production in Britain. Philos Trans R Soc Lond B 281:277–294

Mooney HA, Gulmon SL (1979) Environmental and evolutionary constraints on the photosynthetic characteristics of higher plants. In: Solbrig OT, Jain S, Johnson GB, Raven PH (eds) Topics in plant population biology. Columbia University Press, New York, pp 316–337

Philippot S, Allirand JM, Chartier M, Gosse G (1991) The role of different daily irradiations on shoot growth and root/shoot ratio in lucerne (*Medicago sativa* L.). Ann Bot 68:329–335

Plenet D (1995) Fonctionnement des cultures de maïs sous contrainte azotée. Détermination et application d'un indice de nutrition. Thèse de Docteur de l'Institut National Polytechnique de Lorraine, Université de, Nancy Nancy, 115 pp

Pons TL, van Rijnberk H, Scheurwater I, van der Werf A (1993) Importance of the gradient in photosynthetically active radiation in a vegetation stand for leaf nitrogen allocation in two monocotyledons. Oecologia 95:416–424

Poorter H (1989) Interspecific variation in relative growth rate: on ecological causes and physiological consequences. In: Lambers H, Cambridge ML, Konings H, Pons TL (eds) Causes and consequences of variation in growth rate and productivity of higher plants. SPS Academic Publishing, The Hague, pp 45–68

Rufty TW, Mac Kown CT, Volk RJ (1989) Effects of altered carbohydrates availability on whole-plant assimilation of $^{15}NO_3^-$. Plant Physiol 89:457–463

Sackville-Hamilton NR, Matthew C, Lemaire G (1995) In defence of the –3/2 boundary rule: a reevaluation of self-thinning concepts and status. Ann Bot 76:569–577

Sage RF, Pearcy RW (1987) The nitrogen use efficiency of C_3 and C_4 plants. II. Leaf nitrogen effects on the gas exchange characteristics of *Chenopodium album* L. and *Amaranthus retroflexus* L. Plant Physiol 84:959–963

Sylvester-Bradley R, Stokes DT, Scott RK (1990) Aphysiological analysis of the diminishing response of winter wheat to applied nitrogen. 2. Evidence. Aspects Appl Biol 25:289–299

Smith H (1982) Light quality, photoperception, and plant strategy. Annu Rev Plant Physiol 33:481–518

Ulrich A (1952) Physiological bases for assessing the nutritional requirements of plants. Annu Rev Plant Physiol 3:207–228

Voss RE, Hanway JJ, Dumanil LC (1970) Relationship between grain yield and leaf N, P and K concentrations for corn and the factors that influence this relationship. Agron J 62:726–728

White J (1981) The allometric interpretation of the self-thinning rule. J Theor Biol 89:475–500

Yoda K, Kira T, Ogawa H, Hozumi H (1963) Intraspecific competition among higher plants. XI. Self-thinning in over-crowded pure stands under cultivated and natural conditions. J Polytechnic Institute (Osaka City University) D 14:107–129

CHAPTER 2

Use of the Nitrogen Nutrition Index for the Analysis of Agronomical Data

G. LEMAIRE[1] and J.M. MEYNARD[2]

Contents

2.1 Introduction

In Chapter 1, the theory showed that a curve relating critical plant N% to crop dry weight could be determined for each type of crop, and that the N status of a crop could be estimated by the nitrogen nutrition index, which relates the actual plant N% to the critical N%. The objective of this first section is to give some examples of the advantage of using such an indicator of crop N status for the analysis and interpretation of agronomical data, derived either from experiments or from on-farm surveys.

[1] INRA Station d'Ecophysiologie des Plantes Fourragères, 86600 Lusignan, France
[2] INRA-INA PG Station d'Agronomie, 78850 Thiverval-Grignon, France

Diagnosis of the Nitrogen Status in Crops
G. Lemaire (ed)

2.2
Use of NNI for an a Posteriori Characterization of Experimental Data

2.2.1
Determination of Crop Potential

In experiments where the objective is the determination of the yield potential of a crop in relation to climatic conditions such as radiation and temperature, it is very important to ensure that no N deficiency occurs in the crop at any period of its growth. This verification is particularly important in network experiments where site by site comparisons have to be made. In many of these studies, the experimenter presumes that N was nonlimiting because "large" amounts of N fertilizers were applied, but this precaution may not be sufficient in some circumstances. Figure 2.1 shows that for the same amount of N fertilizer applied, 200 kg ha^{-1}, the values of NNI were near the optimum during the first growth period of a forage sorghum crop, while they decreased very rapidly during the period of regrowth, despite a further application of 200 kgN ha^{-1} after cutting. It can be observed that the application of 200 kgN ha^{-1} after the cut was sufficient only to maintain the crop at a similar N status to the nonfertilized treatment in the

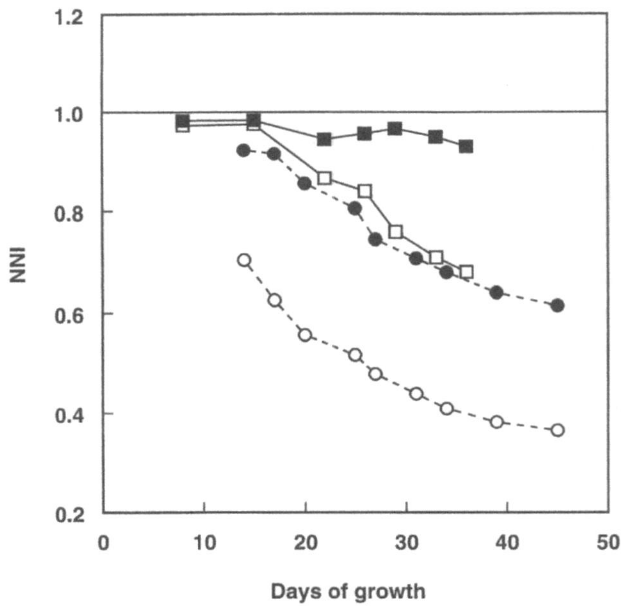

Fig. 2.1. Time course of the nitrogen nutrition index (NNI) of a forage sorghum crop during growth after sowing (■, □) and during regrowth after the firs cut (●, ○); crops were grown in absence of N application (*open symbols*), or with dressings of 200 kg N ha^{-1} applied after sowing and after the cut (*dark symbols*)

previous growth period. In this example, without a diagnosis of crop N status, the differences in crop growth between the two periods with a high level of N application would be attributed to other unknown factors, leading the experimenter to formulate an invalid hypothesis to explain the lower growth potential after cutting.

2.2.2
Crop Response to N Deficiency

In experiments on the effect of N fertilization on crop production, determination of the NNI should provide a useful procedure for the estimation of the range of N deficiency which has been studied. The data presented in Fig. 2.1 show that the range of N supply for a crop can be very different from the range of N application rates. The value of NNI reached by a nonfertilized crop gives an indication of soil N supply, and allows the comparison of crop responses to N fertilizer supply obtained under different soil and climatic conditions. Examples of use of NNI for this purpose are given for grasslands and maize in the following sections.

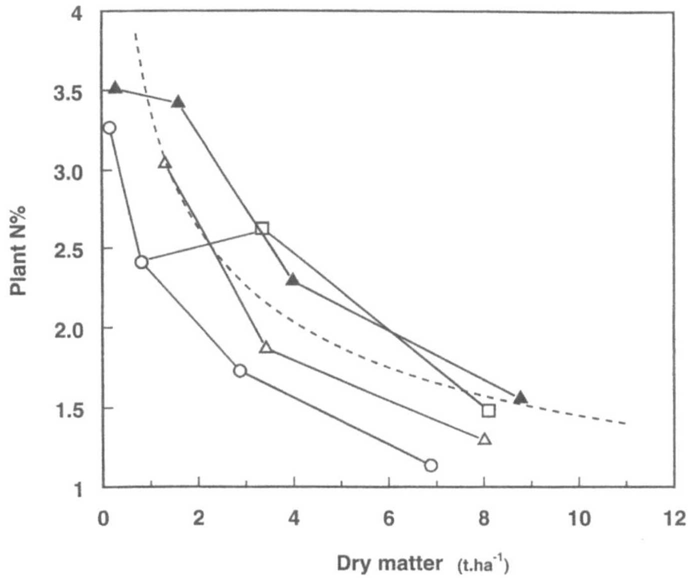

Fig. 2.2. Determination of periods of N deficiency means of the critical plant N% curve for *Penisetum typhoïdes* in Senegal. (After Diouf 1990). The *dotted line* represents the critical plant N% for C_4 plants: N% = $3.4(W)^{-0.37}$. ▲ No N deficiency (two N application of 60 kg N ha^{-1} each); □ N deficiency during vegetative period only (late N application of 60 kg N ha^{-1}); △ N deficiency during reproductive period only (one early N application of 60 kg N ha^{-1}); ○ permanent N deficiency (no N application)

In some experiments, the objective is to determine the effect of a period of N deficiency at a determined stage of the growth cycle, to identify the periods of greatest sensitivity of the plants to N shortage. In these experiments, different fertilizer N dressings are applied in order to create periods of limiting N nutrition. The use of sequential determinations of NNI is therefore necessary to verify a posteriori that the different N fertilizer treatments did indeed lead to the expected timing of N deficiency. An example of the use of NNI for this purpose is given in Fig. 2.2 for a pearl millet crop.

2.2.3 Analysis of Experimental Networks

The use of the NNI as an indicator of crop N nutrition status should allow the classification of observed data according to the degree of N deficiency and the interpretation of other crop growth parameters with the minimum of interference by N nutrition status. Meynard (1985) used the NNI on an experiment network with wheat in northern and western regions of France to identify a group of 56 situations of nonlimiting N supply (NNI greater than or equal to 1) until the begining of stem elongation stage. Table 2.1 shows that in these situations, when the crop biomass remains lower than $50\,g\,m^{-2}$, the majority of plants show a tillering pattern which corresponds to that predicted by the tillering model of Masle and Sebillotte (1981). When the crop biomass is greater than $70\,g\,m^{-2}$, a lower number of tillers can be observed in comparison to the model. The use of the NNI to restrict the analysis of the tillering pattern to situations where one can be sure that no N deficiency occurred, allows the experimenter to assume that the cessation of tillering is due to competition for light when the aboveground biomass reaches 50 to $70\,g\,m^{-2}$, which corresponds to an LAI of 0.85 to 1.20. Without using the NNI, this limitation of tillering by LAI for would not have been determined with precision because of the interference of N deficiency in the cessation of tillering.

Table 2.1. Cessation of tillering and aboveground biomass of 56 wheat crops that were identified as nonlimited by N (NNI ≥ 1) out of 120 observed situations. The cessation of tillering is defined as being when more than 50% of observed plants showed a lack of new tiller appearance in comparison with the model of Masle and Sebillotte (1981)

Aboveground biomass $g\,m^{-2}$	Continuation of tillering	Cessation of tillering
<50	5	0
≥50 and ≤70	6	8
>70	0	37

2.3 Analysis of the Effect of Soil Conditions on Plant N Nutrition

The availability of N for plants is determined not only by the level of N supply, but also by soil conditions, which determine root growth and root activity for mineral N absorption. Soil water deficit and soil structure are two main soil factors which can limit plant N nutrition.

2.3.1 Effects of Drought on Plant N Nutrition

The results of Garwood and Williams (1967a,b) demonstrated that when the upper layer of the soil is dried, the growth of grasses can be impaired by reduction of the absoption of N and other minerals, while the water consumption of the stand is maintained at the optimum level by absorption of water from deeper soil layers. Thus, under dry conditions, the growth of plants can be reduced by a drought-induced N deficiency before a water shortage occurs, which leads to an overestimation of the effect of water deficit per se on crop productivity. Lemaire and Denoix (1987) used the relationship between plant N% and biomass of a tall fescue sward to discriminate situations where water deficit was associated with induced N deficiency from situations where the reduction of growth resulted directly from the water stress itself. Figure 2.3 shows that in 1979 the water deficit did not change the N nutrition status, while in 1985 the N nutrition status of the nonirrigated treatments was lowered in comparison with irrigated conditions. In 1979, the dry period was interupted by rainy episodes which helped to maintain the availability of N concentrated in the upper layers of soil. In 1985, the soil was dessicated to a greater depth, and plants were forced to absorb water from deeper soil layers while N was concentrated mainly towards the surface. These two situations led to very contrasting water use efficiency: when plants consumed water preferentially from the upper soil layers where the N concentration was high, the water use efficiency was similar in irrigated and in nonirrigated conditions, while when plants were obliged to consume water from deeper soil layers, where the N concentration was low, the water use efficiency was depressed as a consequence of N shortage induced by drought (Lemaire and Denoix 1987).

Figure 2.4 shows a reinterpretation of the data of Vough and Marten (1971) on the effect of water deficit on the N% of lucerne crops. Due to the absence of a decrease in plant N% with increase in water deficit, these authors concluded that the N nutrition of lucerne was unaffected by drought. However, as shown in Fig. 2.4, the use of the critical plant N% vs. crop dry weight relationship established by Lemaire et al. (1985) for lucerne shows that the constancy of plant N% is the result of two opposite effects: (1) an increase in critical plant N% as a consequence of the reduction in crop mass by water stress, and (2) a decrease in

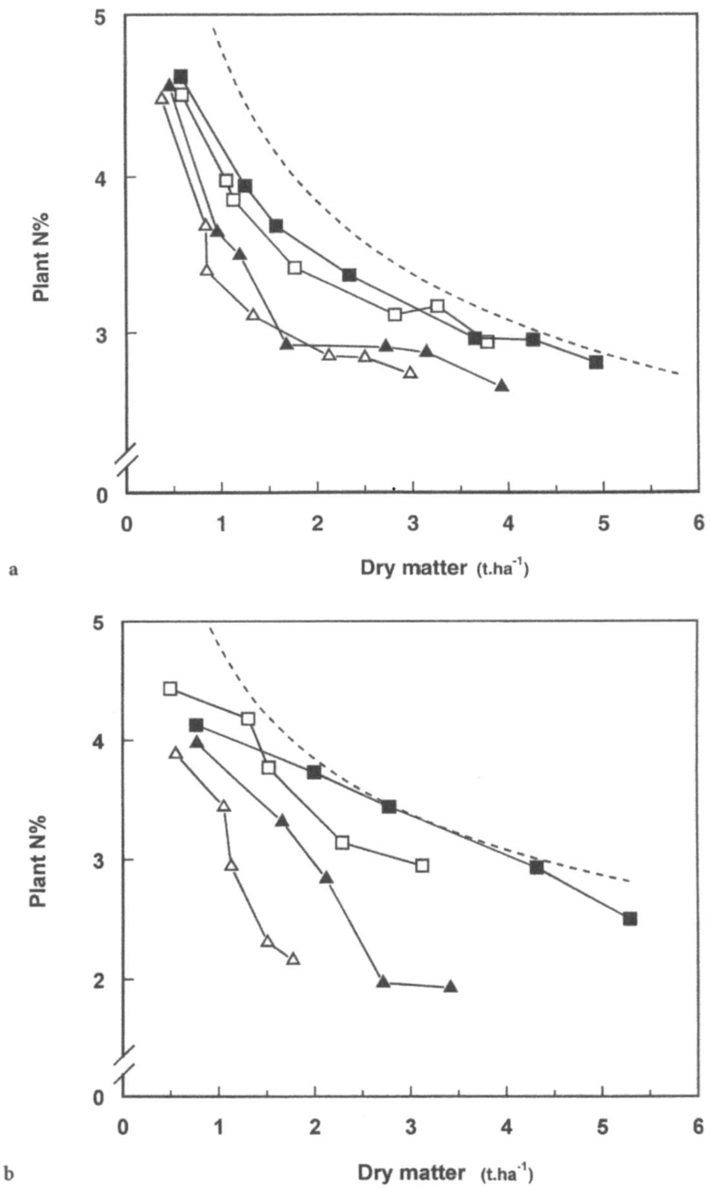

Fig. 2.3a,b. Relationships between plant N% and sward biomass during summer regrowth of tall fescue in 1979 (**a**) and 1985 (**b**) for irrigated (*dark symbols*) and nonirrigated (*open symbols*) plots, and with an application of 60 (▲, △) or 120 (■, □) kg N ha^{-1} after the cut. The *dotted line* represents the critical plant N% curve determined by Salette and Lemaire (1984) for grasses: N% = $4.8(W)^{-0.32}$

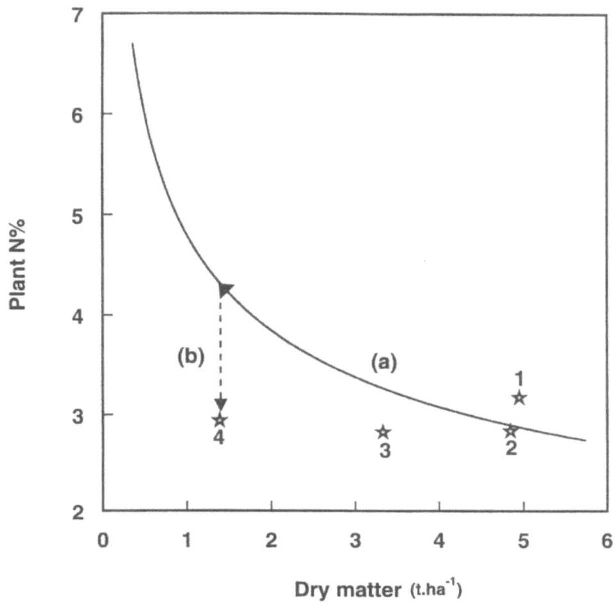

Fig. 2.4. Representation of the effect of water deficiency on plant N% in lucerne. The *line* represents the critical plant N% curve $N\% = 4.8(W)^{-0.32}$. The data are recalculated from Vough and Marten (1971); the *numbers* refer to increasing water stress on a scale from 1 to 4. *a* Theoretical increase in plant N% as a consequence of the reduction in lucerne biomass due to water deficit; *b* the effect of water stress on the N nutrition of the plants due to water stress

the N nutrition status of plants as a consequence of the direct effect of soil water deficit on the N_2 fixation rate.

2.3.2 Effect of Soil Structure

The compaction of soil can disturb many processes involved in crop productivity: decomposition of organic matter, nitrification of ammonium, growth and respiration of roots, and the absorption of water and minerals. These effects can result in a modification of the N nutrition of plants, which can be estimated using the NNI. Thus, Gede-Wibawa (1992) compared the N status of a spring barley crop cultivated with two contrasting soil structures. The first soil was characterized by a fragmented structure favorable to root growth and nitrification, while the second was characterized by heavy wheel compaction on more than 50% of the volume of the arable soil layer. Table 2.2 shows that for the two soil structures, increase in the rate of N fertilizer application increased the level of N nutrition of the crop. Soil compaction decreased the NNI; the N application of 200 kg ha^{-1} was optimum only

Table 2.2. Effect of soil structure on the N nutrition of malting barley as indicated by the crop NNI. (Gede-Wibawa 1992)

N application rate $kg\,ha^{-1}$	NNI at flowering	
	Favorable soil structure	Compacted soil structure
100	0.67	0.59
200	1.00	0.72

with a favorable soil structure, while a greater N application would have been necessary to reach the optimum N nutrition status under conditions of soil compaction.

2.4 Comparison Between Species and Cultivars

The absence of clear differences in the critical plant N% vs. crop dry weight relationship for different species of the same metabolic group during the vege-

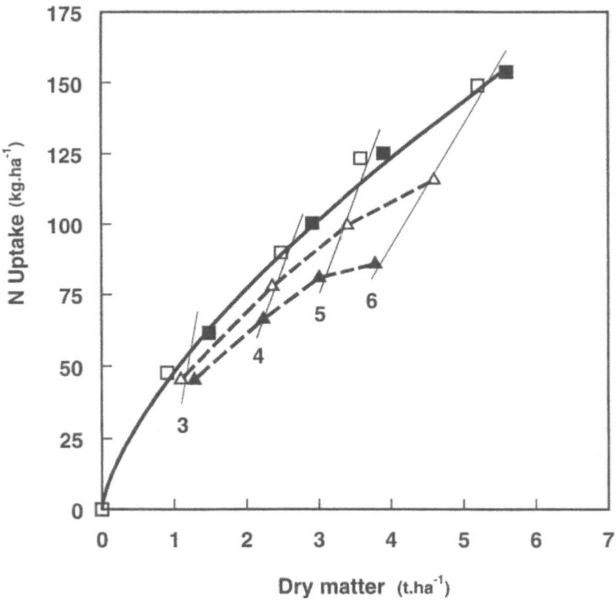

Fig. 2.5. Dynamics of N uptake in relation to dry matter accumulation for tall fescue (*dark symbols*) and cocksfoot (*open symbols*) after an application of 60 (▲, △) or 120 (■, □) $kg\,N\,ha^{-1}$. The *line* represents the critical N uptake curve determined by Lemaire and Salette (1984) for grasses: $N = 48(W)^{0.68}$. The *numerals* refer to weeks of regrowth

tative stage of growth (i.e., excluding grain or tuber filling periods), suggests that it is likely that no significant differences could be found between cultivars of the same species. Lemaire et al. (1985) showed that the differences in plant N% frequently observed by plant breeders between lucerne genotypes were entirely explainable by differences in growth capacity. Therefore, breeding efforts for high plant N% should inevitably be accompanied by a decrease in yielding ability.

Figure 2.5 shows that use of critical plant uptake vs. crop dry weight curves allows the identification of differences in N uptake capacity between different grass species. At optimum N supply, tall fescue and cocksfoot follow the same general relationship; but for a suboptimal level of N supply, cockfoot shows a greater capacity for N uptake.

2.5 Crop Management Studies

When new methods of crop management are tested in experiments, it is important to analyze the causes of the results obtained in order to determine their representativity. Examples of the use of NNI for management of fertilizer N application in wheat were reported by Justes (1993). Table 2.3 shows the test of a new management system for wheat crops adapted for farms with a very low manpower per hectare, where the farmers are sometimes obliged to delay fertilizer applications if this conflicts with sowing of spring crops. This delay induces temporary N deficiency and a deficit in grain yield. In these circumstances, a reduction of the crop density could reduce the intensity of the N deficiency experienced by the crop, as shown by the NNI, and therefore increase the yield in comparison to the normal crop density (Meynard and Aubry 1988).

Table 2.3. Comparison of two crop management systems for wheat crops for situations where farmers are obliged to delay the N application in spring

	Low sowing density	Usual sowing density
Plant density, m^{-2}	240	390
Biomass at the beginning of stem elongation, $g\,m^{-2}$	64	78
Biomass at 1-node stage, $g\,m^{-2}$	122	141
NNI at the beginning of stem elongation	1.10	0.97
NNI at 1-node stage	0.78	0.61
No. of ears m^{-2}	459	448
No. of grains, $m^{-2} \times 10^3$	16.06	14.48
Yield, $t\,ha^{-1}$	7.68	6.93

Table 2.4. Classification of productivity of tall fescue swards on farm plots in relation to sward nitrogen and phosphorus status. N status is determined by NNI; P status is determined by plant P% : N% ratio. (David and Meynard 1990)

P status	N status	Growth rate in kg dry matter per ha and per degree-day		
		8 to 10	6 to 8	<6
P/N ≥ 0.12	≥0.9	1	0	0
	<0.9	5	1	1
P/N < 0.12	≥0.9	0	3	0
	<0.9	0	1	3

2.6 Analysis of Variation in Crop Yield on Farm Networks

Each year, farmers and advisers evaluate and compare the observed yields, and try to explain the observed differences; but in the absence of valuable diagnostic tools, the judgements are very subjective, and cannot easily be used for the improvement of the cropping system. Meynard and David (1992) and Doré et al. (1996) formalized a methodology of yield analysis based on diagnosis of the limiting factors for biomass and yield components. The determination of NNI at the key stages of crops could provide a useful tool for such a diagnosis. Table 2.4 gives an example of use of NNI as a diagnostic tool on farmers' fields to interpret the variation in spring growth of meadows in marshlands in the west of France (David and Meynard 1990). It can be shown that meadows with a low growth rate are deficient in N (NNI < 0.9) and in phosphorus (P/N < 0.12), and that meadows with higher growth rates are never deficient in P, but can often be deficient in N. Ney et al. (Chap. 6 this Vol.) give another example of the use of NNI for diagnosis of limitation by N in pea crops at the farm level, showing that deficient N nutrition appears to be the first factor which limits the yield of crops, because the development of nodules and their activity are frequently limited by parasites (weevils) and soil structure.

2.7 Conclusion

The different examples outlined above show that the use of NNI for interpreting agronomical data can help the agronomist to detect situations where N nutrition is directly or indirectly involved in the reduction of crop yields from their potential. The determination of NNI of crops at the key stages of crop development and just before N fertilizer applications should become a routine tool of diagnosis in field experiments, just like the routine determination of crop LAI, biomass, or yield components. The use of this tool could allow unexpected N effects caused by some experimental treatments to be detected and quantified, and should avoid some misinterpretation of experimental results.

References

David G, Meynard JM (1990) Diagnostic sur les effets des systèmes de culture et des modes de contrôle de l'eau sur prairie de fétuque élevée dans les Marais de l'Ouest. CR Acad Agric Fr 76:108–114

Diouf M (1990) Analyse de l'élaboration du rendement du mil (*Pennisetum typhoïdes* Stapf et Hubb.). Mise au point d'une méthode de diagnostic en parcelles paysannes. Thèse de Doctorat, INA Paris-Grignon, 227 pp

Doré T, Sébillotte M, Meynard JM (1996) A diagnosis method for assessing regional variations in yield. Agric Syst (in press)

Garwood EA, Williams TE (1967a) Soil water use and growth of a grass sward. J Agric Sci Camb 68:281–292

Garwood EA, Williams TE (1967b) Growth water use and nutrient uptake from subsoil by grass swards. J Agric Sci Camb 69:125–130

Gede-Wibawa J (1992) Approche par enquête et expérimentation de l'effet de l'état structural du sol sur la nutrition azotée et l'élaboration du rendement de l'orge de brasserie. Thèse de Doctorat, NIA Paris-Grignon, 235 pp

Justes E (1993) Contrôle de la nutrition azotée du blé par l'analyse des teneurs en nitrate dans la plante: intérêt pour l'ajustement de la fertilisation en milieu à forte fourniture azotée. Thèse de Doctorat, INA Paris-Grignon, 234 pp

Lemaire G, Denoix A (1987) Croissance estivale en matière sèche de peuplement de fétuque élevée (*Festuca arundinacea* Schreb.) et de dactyle (*Dactylis glomerata* L.) dans l'Ouest de la France. 1 Interaction entre les niveaux d'alimentation hydrique et de nutrition azotée. Agronomie 7:381–389

Lemaire G, Salette J (1984) Relation entre dynamique de Coissance et dynamique de prélèvement d'azote pour un peuplement de gramiries fourrageies. I. Etude de l'effet du milien. Agronomie 4:423–430

Lemaire G, Cruz P, Gosse G, Chartier M (1985) Etude des relations entre la dynamique de prélèvement d'azote et la dynamique de croissance en matière sèche d'un peuplement de luzerne (*Medicago sativa* L.). Agronomie 5:685–692

Masle J, Sébillotte M (1981) Etude de l'hétérogénéité d'un peuplement de blé: origine des differentes Catègories d'individus du peuplement, èléments de description de sa structure. Agronomie 1:217–223

Masle-Meynard J (1981) Elaboration du nombre d'épis d'un peuplement de blé d'hiver en situation de compétition pour l'azote. 1 Mise en évidence d'un stade critique par rapport à la montée d'une talle. Agronomie 1:623–632

Meynard JM (1985) Les besoins en azote du blé d'hiver jusqu'au début de la montaison. Agronomie 5:579–689

Meynard JM, Aubry C (1988) Itinéraires techniques pour le blé en conditions d'excès d'eau. Perspect Agric 126:80–89

Meynard JM, David G (1992) Diagnostic sur l'élaboration du rendement des cultures. Cah Agric 1:9–19

Vough LR, Marten GC (1971) The influence of soil moisture and ambient temperature on yield and quality of alfalfa forage. Agron J 63:40–42

PART II

The Nitrogen Requirement of Major Agricultural Crops

CHAPTER 3

Grasslands

M. Duru[1], G. Lemaire[2], and P. Cruz[3]

Contents

3.1 Introduction

Natural grasslands have a very large range of diversity both in terms of botanical composition, and with respect to edaphic and climatic conditions, in temperate and tropical regions. Moreover, grasslands are subjected to a wide variety of defoliation treatments according to management plans. This diversity of conditions leads to large temporal and spatial variations in sward structure: i.e., botanical composition, plant morphology and architecture, plant growth habits, which make it difficult to use a diagnosis of the sward state to predict either its growth potential or its ability to be grazed and to provide adequate feed for animals.

In addition, mineral deficiency (P and K) and/or water deficit interact with plant N nutrition in many circumstances, particularly in natural grasslands. Therefore, in order to use the theoretical concept of critical plant N% for diagnostic purposes in grasslands, several questions have first to be answered:

1. Could the large differences is sward structure due to differences in botanical composition, plant morphology and plant architecture lead to variations in the pattern of decline in critical plant N% in relation to biomass accumulation of the sward for different types of grasslands?

[1] INRA Station d'Agronomie, Auzeville BP 27, 31326 Castanet-Tolosan, France
[2] INRA Station d'Ecophysiologie des Plantes Fourragères, 86600 Lusignan, France
[3] Station d'Agropédoclimatologie INRA, Petit-Bourg BP 1232, 97185 Pointe à Pitre Cédex, Antilles Françaises. Present address: INRA Station d'Agronomie, BP 27, 31326 Castanet-Tolosan Cédex, France

Diagnosis of the Nitrogen Status in Crops
G. Lemaire (ed)

2. Can the other limiting factors, such as water deficit or P and K deficiency, affect the relationship between critical plant N% and sward biomass, and how can these influences be taken into account for diagnostic purposes?
3. How can the nitrogen nutrition index be used for diagnostic procedures under natural conditions?

Each of these questions is analyzed in a specific section below. Some of them can be answered directly from the data we have obtained, and it may be possible to propose a generalized diagnostic procedure for grasslands. Other questions require new experiments in order to develop more suitable diagnostic procedures for specific situations.

3.2
Critical Plant N% and Sward Biomass in Different Types of Grassland

The data presented in Chapter 1 concerned mainly pure grass or legume stands. We have shown in that chapter that the empirical relationship critical plant N% vs. sward biomass appeared to be very similar for different species. However, the range of grass species used for this comparison (see Chap. 1, Fig. 1.2) was very limited, and concerned only cultivated species (tall fescue, cocksfoot, perennial ryegrass) having a similar architecture (erected tillers). We have no direct information on the influence of very contrasting sward structures obtained with prostrate or stoloniferous species, as opposed to erected ones. However, even with a very different stand structure (horizontal leaves), lucerne has a similar critical plant N% vs. sward mass relationship to erect grasses. Moreover, a pure sward of white clover shows a critical plant N% decline similar to that of lucerne or erect

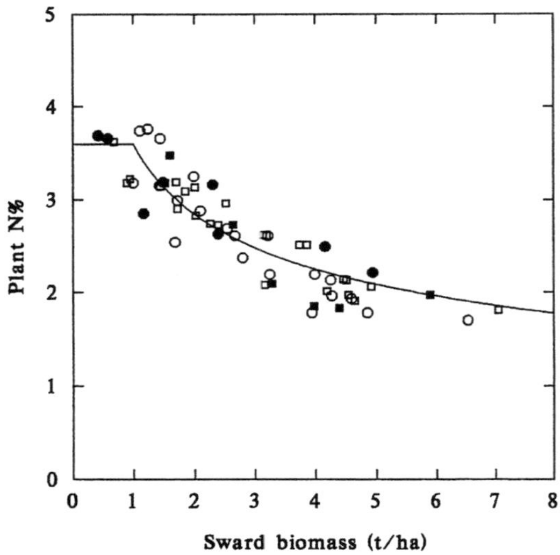

Fig. 3.1. Plant N% vs. aerial biomass for C_4 tropical grasses. *Bracharia* ●; *Digitaria* ○; *Panicum* ■; *Dichantium* □, with high N supply. Reference curve established on erect C_4 species: N% = $3.6\,DM^{-0.34}$

grasses, despite its horizontal leaves and its stoloniferous growth habit (Soussana and Arregui 1995).

For C_4 grasses, Fig. 3.1 shows that different species such as *Panicum maximum*, *Digitaria decumbens*, *Dichantium aristatum*, or *Brachiaria decumbens*, with contrasting morphologies and growth habits, show a similar pattern of decline of critical plant N% with increasing sward biomass. Moreover, this decline in critical plant N% is similar to that of the maize or sorghum crops described in Chapter 1, despite the very different structure of these different stands.

Thus, the conclusion of Chapter 1 on the absence of differences in the decline of critical plant N% between species of the same metabolic group growing in dense stands has clearly been confirmed. If no (or small) differences can be observed between pure stands of species of contrasted morphology, we can suppose that dense stands of mixed species such as natural grasslands will follow the same general relationship, whatever their botanical composition.

In order to verify this assumption, we have constructed a data base containing the results of several experiments on a large range of natural grasslands, including swards composed of more than 50% of nonlegume dicotyledons and noncultivated grasses like *Festuca rubra* or *Anthoxanthum odoratum*. These swards were located in three different sites in mountainous regions of the southwest of France: the Central Pyrenees at an altitude of 1250 m, Aubrac at an altitude of 900–1000 m, and the Aveyron Valley at an altitude of 600–800 m. For each type of sward we had two N application rates, N_0 with no N applied, and N_1 with 150 kg ha^{-1} applied at the end of winter, each sward had been subjected to different modes of organic fertilization during the 5 preceding years (Duru et al. 1994). During spring growth after the different applications of N, several successive cuts were performed to determine the time course of sward biomass and plant N% for each N treatment and each type of grassland. For each date of cutting, the difference in sward biomass between the two N treatments was tested:

- If N_1 biomass $>N_0$ biomass, only the data points corresponding to N_0 were considered to be the limiting N condition. Because it was impossible to have any information on whether the N_1 treatment was limiting or not, the data point corresponding to N_1 was rejected from the data base.
- If $N_1 = N_0$, the two data points were introduced into the data base as a nonlimiting N condition.

Figure 3.2 shows that the general equation for C_3 grasses described in Chapter 1, that is $N_c\% = 4.8(W)^{-0.32}$, discriminates the limiting N conditions from the nonlimiting ones. Only one pair of data points seemed to have been classified as limiting instead of nonlimiting. Nevertheless, these corresponded to a relatively large sward biomass greater than 8000 kg DM ha^{-1}, and at this stage of growth, and for this type of sward, the accumulation of dead plant material resulted in a reduced plant N% (Duru 1994).

In order to verify the lack of any effect of botanical composition on evolution of plant N% for the sward, we selected, from the same site, the two most contrasting pastures in terms of botanical composition. Figure 3.3 shows that for a similar

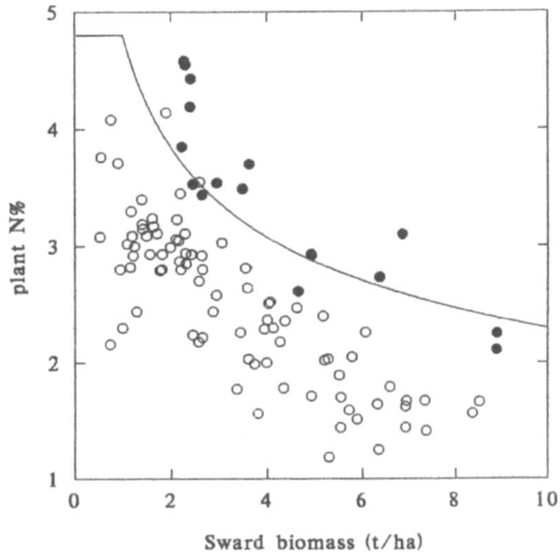

Fig. 3.2. Plant N% for different natural grasslands in nonlimiting (●) or in limiting N supply conditions (○); — critical N% for C_3 grasses; $N\% = 4.8\,DM^{-0.32}$

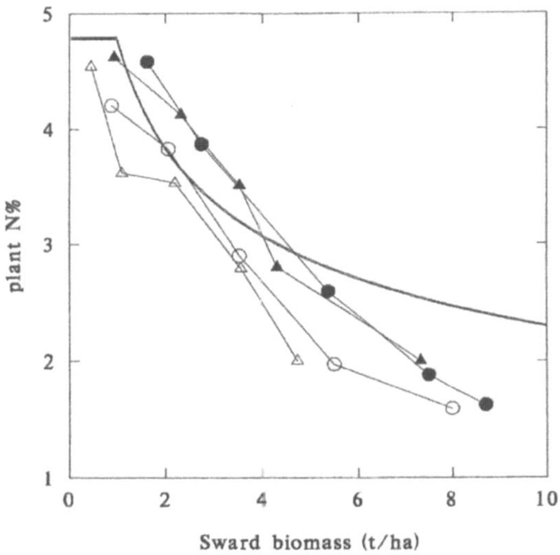

Fig. 3.3. Time course of N% during spring growth of two different pastures of contrasting botanical composition. Pasture A (● ○) is dominated by *Dactylis glomerata* and *Chaerophyllum aureum*, pasture B (▲ △) is dominated by *Festuca rubra* and *Sanguisorba minor*. Samples were taken after N application of $120\,kg\,ha^{-1}$ in pastures given P fertilizer applications during the 3 preceding years (*closed symbols*) or grown without added P (*open symbols*). The regression line represents the reference curve of critical plant N% for C_3 species

application rate of $150\,kg\,N\,ha^{-1}$ and with a high level of P application, the decline in plant N% with increasing sward biomass was similar for the two pastures. Nevertheless, the difference in botanical composition between the two swards led to a difference in biomass accumulation, and therefore a difference in plant

N% at a given date of harvest. Furthermore, Fig. 3.3 shows that the single application of 150 kg N ha^{-1} was not enough to maintain both pastures in nonlimiting N conditions for the whole period of growth considered. This application rate was only enough to achieve a biomass of about 5000 kg DM ha^{-1} without any N limitation. After this point, the nitrogen nutrition index of the two swards become less than 1 and decreased rapidly, indicating a progressively more intense N deficiency towards the end of the growing period. We may observe here that in the absence of P application, the level of N nutrition was depressed for both pastures. This aspect of the interaction between N and P will be analyzed in the next section.

The absence of an effect of botanical composition on the plant N% vs. sward biomass relationship must be analyzed in the light of the theory developed in Chapter 1, where we indicated that the decrease in critical plant N% of a crop parallels the decrease in the leaf area ratio (LAR). If this assumption is true, it implies that due to the absence of a clear difference in the plant N% decline pattern between species with different morphologies, the decline in LAR with increasing sward mass is similar for a large range of species and is therefore not affected by the botanical composition of the grassland. The LAR of swards is rarely measured; however, the leaf/stem ratio (L/S) is sometimes determined as an estimator of herbage quality and as an indicator of sward structure. It is clear that, for natural grasslands, the leaf/stem ratio varies greatly in the spring according to the pasture composition (vegetative vs. reproductive states of the grass tillers, botanical composition). Moreover, for spring growth, there is a unique relationship between the leaf weight ration (LWR) and biomass for a large range of botanical compositions and herbage nutrient statues, whatever the sampling date (Duru 1994). As with late

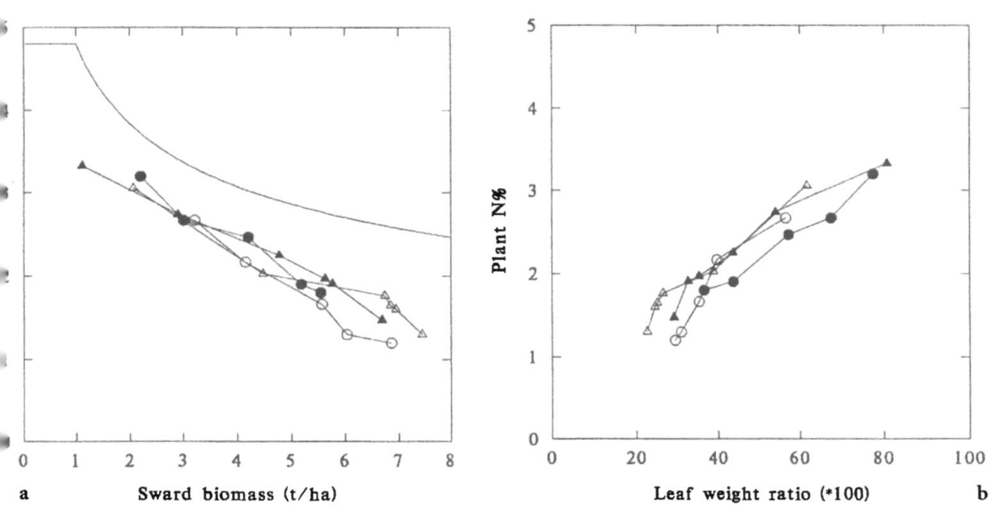

Fig. 3.4a,b. Plant N% in relation to sward biomass (**a**) or leaf weight ratio (**b**) during spring growth of two different pastures of contrasting botanical composition (see Fig. 3.3). **a** ● ○; **b** ▲ △ after N application of 120 kg ha^{-1} (*closed symbols*) or P application of 50 kg ha^{-1} (*open symbols*)

sampling dates of spring growths (after the heading stage of grasses), the N concentration is greater for green leaf than for stem, and the nitrogen concentration of the whole herbage depends greatly on the leaf/stem ratio. This is why, for different pastures having the same NNI (Fig. 3.4a), there is a positive relationship between N concentration and the percentage of leaf in the biomass, whatever their botanical composition (Fig. 3.4b). However, this relationship has no general value for earlier growth stages or for vegetative regrowths. In these cases, it might be impossible to calculate a leaf/stem ratio, or this criterion might not be a good estimator of the leaf area ratio (LAR) because of large variations in the specific leaf weight (SLW) of grass species (Poorter and Remkes 1990; Duru et al. 1994). As shown in Chapter 1 for sorghum, leaf mass does not exactly represent the metabolic component of the plant because it includes a greater proportion of structural component as the leaf area increases.

We have verified that a multispecies sward of grass species of contrasting morphology and dicotyledon species (legumes and nonlegumes) followed the general model developed in Chapter 1. Such a sward was maintained in nonlimiting N conditions by applying $150\,kg\,N\,ha^{-1}$ at the end of winter. The accumulation of biomass was recorded by successive cuttings during spring growth. The LAI of the sward and the plant N% were recorded simultaneously. These measurements were continued for three successive years. Figure 3.5a shows the decrease of plant N% with increase in sward biomass. The data points are equally distributed on either side of the line of the general critical plant N% vs. sward biomass equation. Thus, the N nutrition of the sward was maintained at a nonlimiting level during the 3 years considered. Figure 3.5b shows that the LAI of the sward increased less rapidly in relative terms than the biomass:

$$LAI = 1.9(W)^{0.726} \qquad R^2 = 0.96.$$

This implies a decrease of the LAR with increasing sward biomass:

$$LAR = LAI/W = 1.9(W)^{-0.274}.$$

The power coefficient of this equation is close to that of the relationship between critical plant N% and sward biomass, that is –0.32. This convergence between the two coefficients leads to a linear relationship between critical plant N% and LAR as represented in Fig. 3.5c:

$$N_c\% = 218.6 LAR + 0.55 \qquad R^2 = 0.79.$$

Thus, the theory developed in Chapter 1, which predicted a linear relationship between critical plant N% and LAR for monospecific stands, is here confirmed for multispecies swards composed of grass and dicotyledon species of contrasting morphologies. A single relationship can therefore be used to describe the decline in critical plant N% with increasing sward biomass for a large range of multispecies grasslands, whatever their botanical composition.

Before a general value of such an index based on this critical plant N% vs. sward biomass relationship can be established it is, however, necessary to demonstrate that it reflects the actual status of the nitrogen nutrition of the sward and is

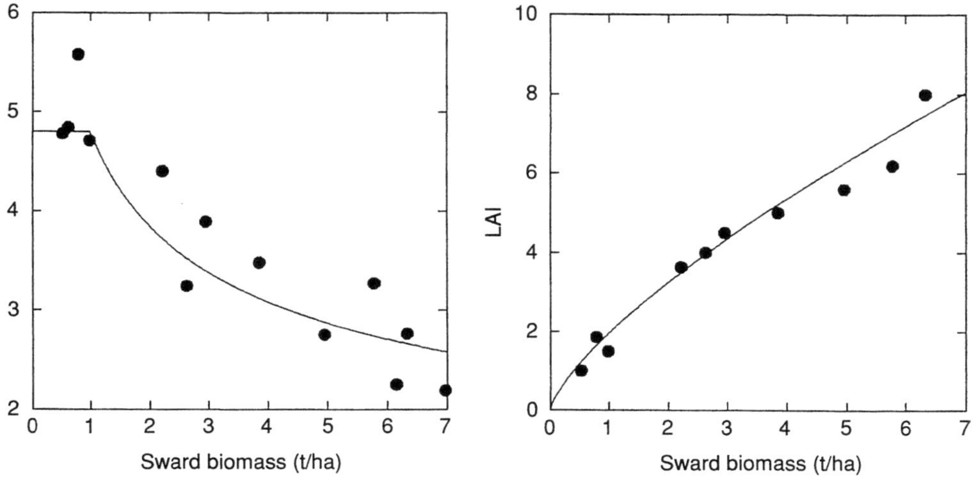

b

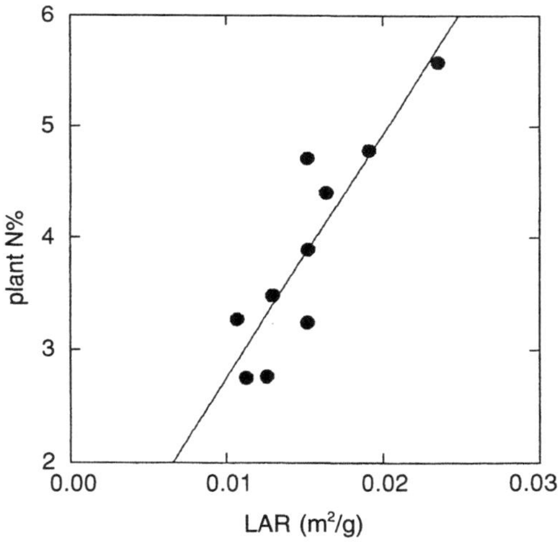

c

Fig. 3.5a–c. Time course of plant N% (**a**) and LAI (**b**) with sward biomass during spring growth for pastures maintained in nonlimiting N conditions by ample N application rates. The relationship between plant N% and leaf area ratio (LAR) is shown in c

not influenced by other environmental factors, such as water deficit or mineral deficiency.

3.3 Interactions with Other Limiting Factors

Among the different environmental factors likely to interact with the N nutrition of the plant, phosphorus (P) must be carefully considered because of the importance of P deficiency in natural grasslands on a world scale and because of its direct synergy or antagonism as an anion with NO_3^-.

The effects of P deficiency on the N nutrition of swards can be studied using the following procedure:

- Estimation of the direct effect of a P deficiency on sward growth processes, which induces an indirect increase in plant N% as a consequence of the reduced biomass accumulation.
- Determination of the direct effect of P on soil N availability through mineralization of organic matter and/or stimulation of root growth (Callot et al. 1982).

To analyze the effect of P deficiency, we used data obtained from two different pastures characterized by low P availability. Two levels of P application ($P_0 = 0$ and $P_1 = 120\,\mathrm{kg\,P\,ha^{-1}}$) were combined with two levels of N application ($N_0 = 0$ and $N_1 = 120\,\mathrm{kg\,N\,ha^{-1}}$). Biomass accumulation was measured four times after N application during the spring growth periods of the 2nd and 3rd years after the P application. Plant N% and P% were determined at each harvest date. The P status of each sward was classified as deficient or satisfactory according to the phosphorus nutrition index method developed by Salette (1982, 1990) and Duru (1992). In the treatment with P deficiency, for an identical value of NNI, we observed that the relative accumulated biomass was lower than for the treatment without P deficiency (Fig. 3.6). The application of P provoked two different effects:

- An increase in the NNI of the sward from about 0.60 to 0.75 for the N_0 treatment and from 0.9 to 1.0 for the N_1 treatment, these increases in NNI leading to corresponding increases in sward biomass accumulation, i.e., an *indirect effect of P application.*
- An increase in the relative accumulated biomass at a similar NNI, which corresponds to the *direct effect of P application.*

Fig. 3.6. Ratio of observed to potential sward biomass vs. the nitrogen nutrition index for two pastures with P supply (*closed symbols*) or without P supply (*open symbols*) and with high (■ □) or low (● ○) N rates; fitted regressions are shown as *dotted lines*

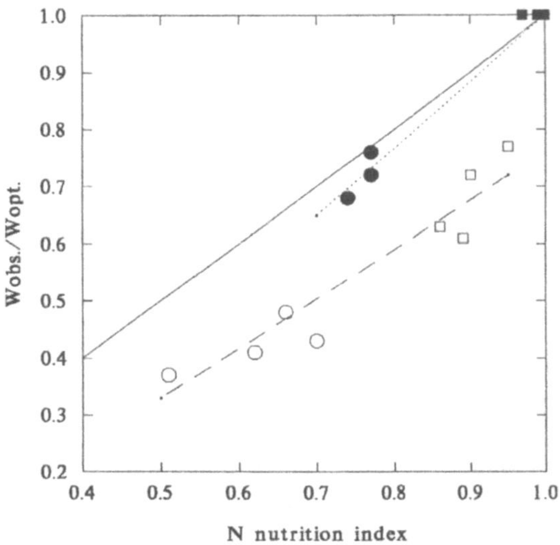

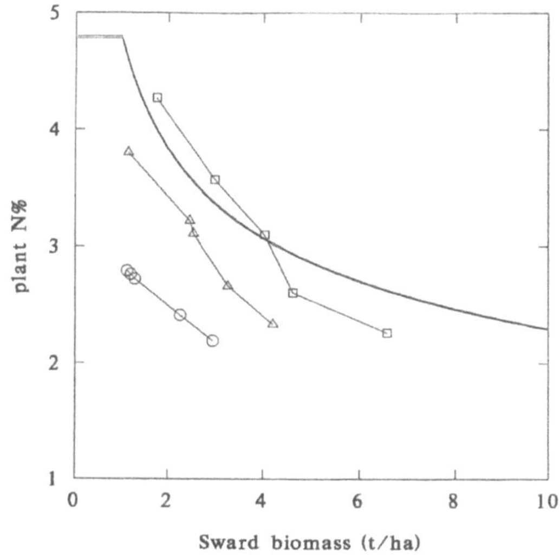

Fig. 3.7. Time course of plant N% during spring growth for pastures which had received: no P fertilizer over the last 5 years (○); an application of P in the preceding year only (△); and an annual P application of 50 kg ha^{-1} over the 5 preceding years (□)

Thus, as shown previously in Fig. 3.3, in situations of P deficiency, the application of P fertilizer increases the availability of N for pastures, as shown by the increase in NNI. Another example of the effect of P supply on the N nutrition status of grasslands is shown in Fig. 3.7. In the absence of any further N application, the NNI of the sward progressively reached the optimum value in response to the accumulated application of P over the previous 5 years.

Thus, the use of the nitrogen nutrition index allows not only the identification of N deficiency due directly to a lack of N supply, but also the determination of situations where the availability of N is depressed by another limiting factor, such as P deficiency. Similar results have also been obtained with K deficiency in natural grasslands (Pons and Lafon 1991; Salette and Huché 1991). Lemaire and Denoix (1987) demonstrated that soil water deficit could also induce a reduction in soil N availability. In these circumstances, as shown in Fig. 3.8, the values of NNI observed in dry conditions are often lower compared to those in irrigated conditions (Duru and Langlet 1994).

3.4 Use of the Nitrogen Nutrition Index of Swards as a Diagnostic Tool

As demonstrated in the preceding sections, the nitrogen nutrition index can be used as an estimator of the N nutrition status of a sward over a large range of botanical compositions and environmental conditions. Nevertheless, all the data which we used for this purpose were obtained in situations of herbage accumulation, i.e., during a regrowth period after cutting or after severe grazing. In these

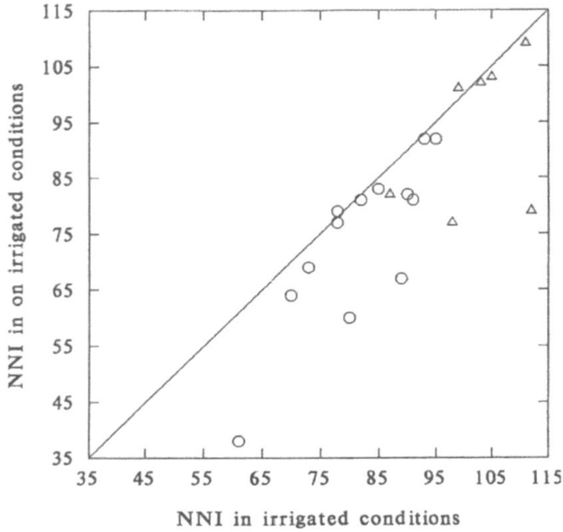

Fig. 3.8. Comparison of nitrogen nutrition index for lucerne (△) and cocksfoot (○) under irrigated and nonirrigated conditions for seven regrowths in the summer

circumstances, the biomass accumulation in swards is a result of increasing plant size and the corresponding decline in critical plant N% is well described by the theory and models developed in Chapter 1. However, when a part of the leaf material is senescing and the dead material accumulates in significant proportions, the above theory cannot be strictly applied, because biomass does not continue to increase at the same rate. Moreover, in continuously stocked swards, the use of the NNI becomes totally irrelevant because the accumulated sward biomass is not directly the result of plant growth, but rather the consequence of a dynamic equilibrium between herbage growth, herbage senescence, and herbage consumption. Since animals constantly remove the upper layer of the canopy, we have no means of directly relating plant N% to sward biomass. Until now we have had no satisfactory procedures for the determination of the N nutrition status of a sward subjected to very frequent defoliations (note here that this is also the case for turf grasses). The only recommended method should be to use exclusion cages for short periods to allow for sufficient sward biomass accumulation in the absence of defoliation, and to postulate that the NNI measured under these cages should represent the N nutrition status of the grazed area during the same period. However, the problem with the determination of the biomass of very short swards is the error due to soil contamination and the difficulties of separating senescent from green plant material, which make the use of such a method under grazing conditions impractical. Thus, we have to accept this drawback and use the NNI as a diagnostic tool only in situations of herbage accumulation following severe defoliation, where the accumulation of senescent plant material is negligible.

Several examples of the use of NNI can be given to illustrate the worth of this diagnostic procedure. The first example concerns a regional diagnosis. Figure 3.9 shows the distribution frequency of the values of NNI recorded in early spring for two regions in France, Aveyron and Aubrac. We may note the variability within each region and the similarity of the frequency distribution between the two regions. In each region, the NNI values are related to the amount of mineral or organic N applied, but the correlation coefficients are low, indicating that soil conditions and the past management of the pastures are more important than recent fertilizer application for determining actual sward N status.

The second example concerns the interpretation of experimental data. Many experiments on grasslands are carried out over several years, and a problem arises from the difficulty in separating annual effects from cumulative (or sward aging) effects.

The example illustrated in Fig. 3.10 concerns the analysis of the response of a tall fescue sward to a given N application rate in spring during the 4 years after sowing. The common observation of a decline in sward productivity following the 1st year after sowing can be well understood by examining Fig. 3.10: the decline in sward productivity with age is due to a decrease in the capacity of the soil to provide N for the sward, and not to a decrease in the growth potential of the sward itself. This is shown by the progressive decline in NNI of the sward during the 4

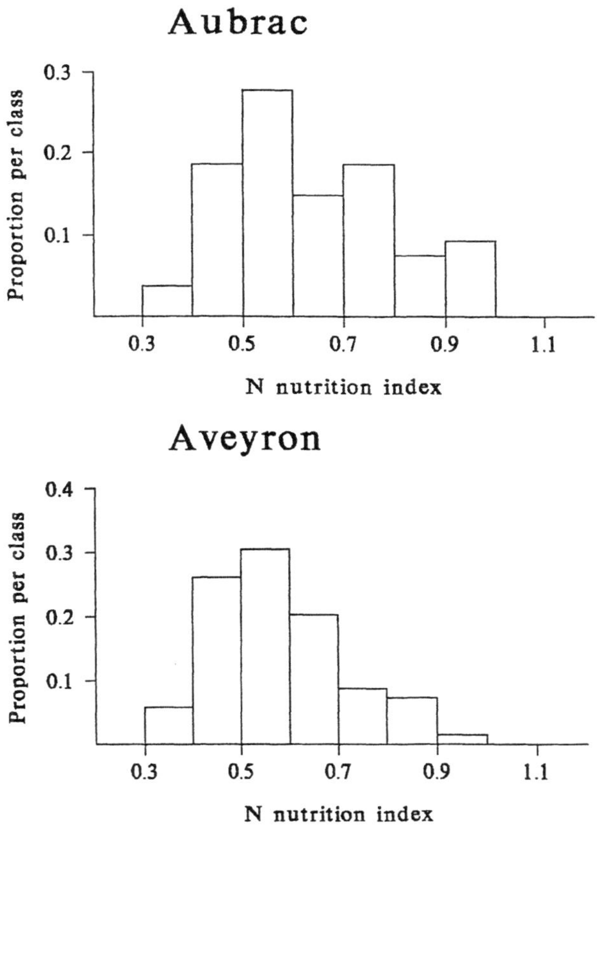

Fig. 3.9. Distribution of nitrogen nutrition index in pastures in two areas in France

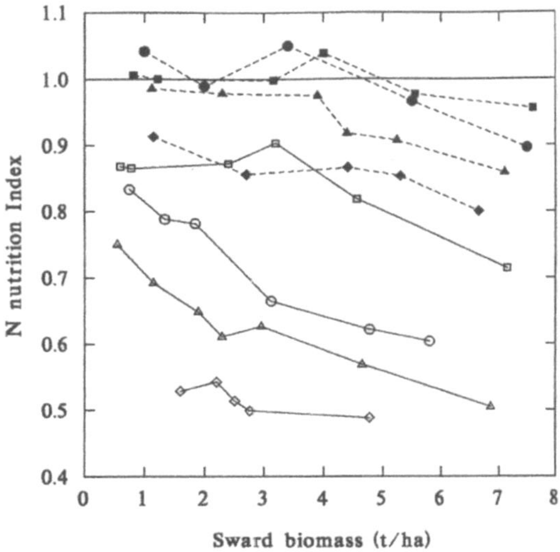

Fig. 3.10. Nitrogen nutrition index vs. sward biomass for tall fescue (cv. Clarine) with no N applied (*open symbols*) or 120 kg ha^{-1} N applied (*closed symbols*) for four consecutive spring growths (□, ○, △, ◇: 1, 2, 3, 4 years after sowing)

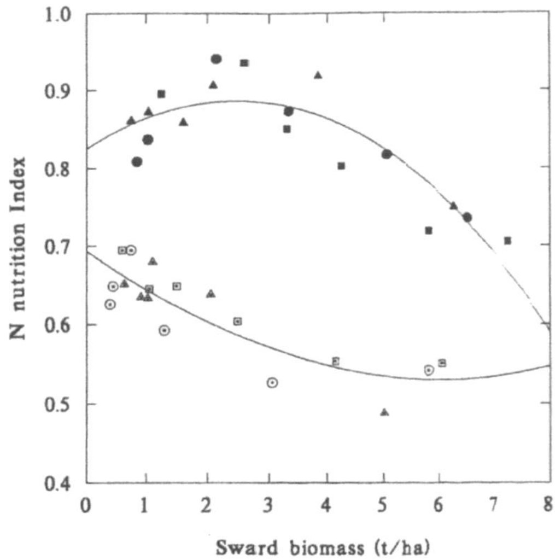

Fig. 3.11. Nitrogen nutrition index vs. sward biomass for permanent pastures with no N applied (*open symbols*) or 100 kg ha^{-1} N applied (*closed symbols*) for three spring growths (three different symbols); — fitted regressions for the two N treatments

years considered, particularly in the N_0 treatment. We may suppose that an increased N application rate during the last 2 years should have allowed the tall fescue sward to reach an optimum N nutrition status. In the absence of any diagnosis of the N nutrition status of the sward, it would have been difficult to detect the true cause of this aging effect, which is commonly attributed to a decrease in the potential productivity of the sward.

By contrast, Fig. 3.11 illustrates the apparent stability of a permanent pasture maintained in equilibrium under a given management regime for several years. In

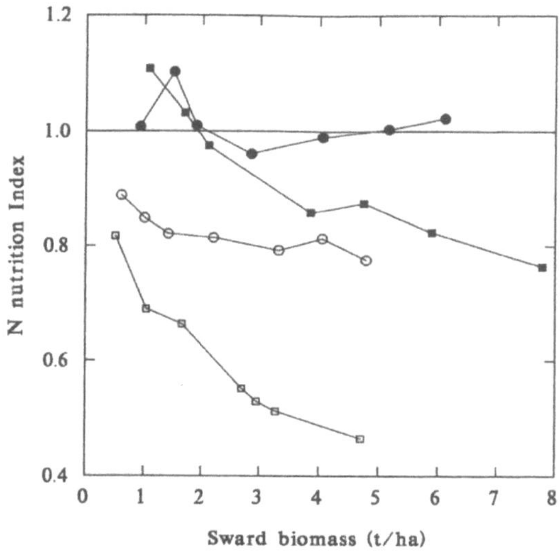

Fig. 3.12. Nitrogen nutrition index vs. sward biomass for permanent pastures with no N applied (*open symbols*) or 100 kg ha^{-1} N applied (*closed symbols*) for two spring growths: ○ ● 1 year after change from grazing to cutting management, or □ ■ 2 years after the change

the absence of N application, similar values of NNI were achieved during the three successive springs, indicating remarkably constant soil N dynamics. With an application of 100 kg N ha^{-1}, the evolution of NNI paralleled that of the N_0 treatment, indicating that the fertilizer N remained available for plants throughout the period considered. Under these conditions, differences observed in sward biomass accumulation between years cannot be attributed to differences of growth caused by variation in soil N availability, but only to differences in climatic conditions that had no apparent effect on soil N dynamics.

This buffered behavior of native grasslands can be observed only if there is no severe perturbation of the soil/sward system. A rapid change in management methods can lead to a rapid modification of the N nutrition status of the sward, as illustrated in Fig. 3.12. Two years after the change from a grazing to a cutting regime, the N nutrition status of this permanent pasture appeared to have been depressed, and a greater N application rate would have been necessary to maintain its productivity level.

3.5 Conclusion

Grasslands have a pattern of critical plant N% decline in relation to sward biomass accumulation similar to that of pure erect grass stands: despite a large range of botanical composition and sward architecture (natural grasslands), or contrasting plant morphologies and growth habits (pure stands). However, as has been shown in Chapter 1, the coefficient a of the curve ($N\% = aW^b$) is different for C_3 and C_4 grasses.

We have demonstrated that factors which limit growth, such as water deficit, or P and K deficiencies in the stand, do not affect the relationship between the critical plant N% and sward biomass. However, very often these factors decrease the availability of soil N for plants, and hence the NNI. This is why the diagnosis of herbage N status should be interpreted carefully in such cases, or the diagnosis may be wrong.

The nitrogen nutrition index is a powerful tool for diagnosis in natural conditions, on a regional level, to compare plots and to assess the effect of some agricultural practices such as past fertilization management. More often it is used as an aid for more effective and accurate grassland experiments, such as the comparison of fertilizer treatments or the assessment of soil N availability and its variability. However, all the data to be used for such a purpose should be obtained after cutting or severe grazing. In the case of continuously stocked pastures, exclusion cages should be used, or more appropriate diagnostic procedures should be developed.

References

Callot G, Chamayou H, Martens C, Salsac L (1982) Les interactions sol-racine. Incidence sur la nutrition minérale. INRA, Paris, 325 pp

Duru M (1992) Diagnostic de la nutrition minérale de prairies permanentes au printemps. I. Etablissement de références. Agronomie 12:219–233

Duru M (1994) Mineral nutrition status botanical composition of pastures. II. Effect on nitrogen concentration and digestibily of herbage. Eur J Agron 3:93–174

Duru M, Langlet A (1994) Croissance de repousses de luzerne et dactyle. I. Interaction eau azote. Agrochimica 38:285–294

Duru M, Balent G, Langlet A (1994) Mineral nutrition status and botanical composition of pastures. I. Effect on herbage accumulation. Eur J Agron 3:43–51

Lemaire G, Denoix A (1987) Croissance estivale en matière sèche de peuplements de fétuque élevée et de dactyle dans l'Ouest de la France. II. Interaction entre les niveaux d'alimentation hydrique et de nutrition azotée. Agronomie 7:381–389

Pons Y, Lafon E (1991) Comment déterminer les facteurs limitants d'une prairie au printemps? Cas des Marais de l'Ouest. Fourrages 125:19–28

Poorter H, Remkes C (1990) Leaf area ratio and net assimilation rate of 24 wild species differing in relative growth rate. Oecologia 83:553–559

Poorter H, Remkes C, Lambers H (1990) Carbon and nitrogen economy of 24 wild species differing in relative growth rate. Plant Physiol 94:621–627

Salette J (1982) The role of fertilizers in improving herbage quality and optimization of its utilization. Proc 12th Int Potash Institute Congr, Goslar, June 1982, 305 pp. Ed IIP, Berne, pp 117–144

Salette J (1990) The effect of level of nitrogen nutrition upon mineral content and removal in grasses and wheat. Fert Res 26:229–235

Salette J, Huché L (1991) Diagnostic de l'état de nutrition minérale d'une prairie par l'analyse minérale du végétal: principes, mise en oeuvre, exemples. Fourrages 125:3–18

Sousana JF, Arregui MC (1995) Impact de l'association sur le niveau de nutrition azotée et la croissance du ray-grass et du trèfle blanc. Agronomie 15:81–96

CHAPTER 4

Wheat, Barley, and Durum Wheat

E. Justes[1], M.H. Jeuffroy[2], and B. Mary[3]

Contents

4.1 Introduction

In the first part of this chapter, we propose to determine the curve of critical plant N% for cereals during the development of the crop ranging from tillering to anthesis, for different cultivars and growth conditions. The results are mainly derived from Justes et al. (1994) with winter wheat. The validity of the curve of critical plant N% will also be evaluated for spring barely and durum wheat.

In the second part, the relationship between the plant N status and yield formation is examined with the nitrogen nutrition index (NNI). The qualities of the NNI for N diagnosis are discussed for cereal crops.

[1] INRA Unité d'Agronomie de Châlons-Reims, 2 Esplanade Roland Garros, BP 224, 51686 Reims Cédex 2, France
[2] INRA-INA PG Laboratoire d'Agronomie, 78850 Thiverval-Grignon, France
[3] INRA Unité d'Agronomie, rue Fernand Christ, 02007 Laon Cédex, France

Diagnosis of the Nitrogen Status in Crops
G. Lemaire (ed)

4.2 Determination of a Critical Plant N% for Cereal Crops

Example of Experimental Curves Relating Plant N% and Crop Biomass. Figure 4.1 gives an example of experimental curves obtained from four N fertilizer treatments on winter wheat (Justes et al. 1994). This indicates that the decrease in plant N% with increasing crop biomass is not systematic: the fertilizer N supplied on April 18, when the plants were at a 1-cm-ear stage of development or Feekes stage 5 (Large 1954), resulted in an increase in the N concentration of the shoots at the onset of elongation (May 2). This was particularly obvious for treatment N_{210}, which had received the highest level of fertilizer N. More regular kinetics were observed after this date because the plants received no other N supply. Generally, and also in this experiment, statistical differences in aerial biomass between experimental treatments appear less rapidly than differences in total N concentration or in the amount of N accumulated in the shoots. This is due to less variability among replicates of total N measurements for shoot dry matter measurements, and also to the fact that N uptake rate varies more between treatments than growth rate. It is therefore difficult to estimate the critical N concentration for a given amount of dry matter with the results of one experiment. This example also illustrates the chronology of an N deficiency appearance: firstly, the N

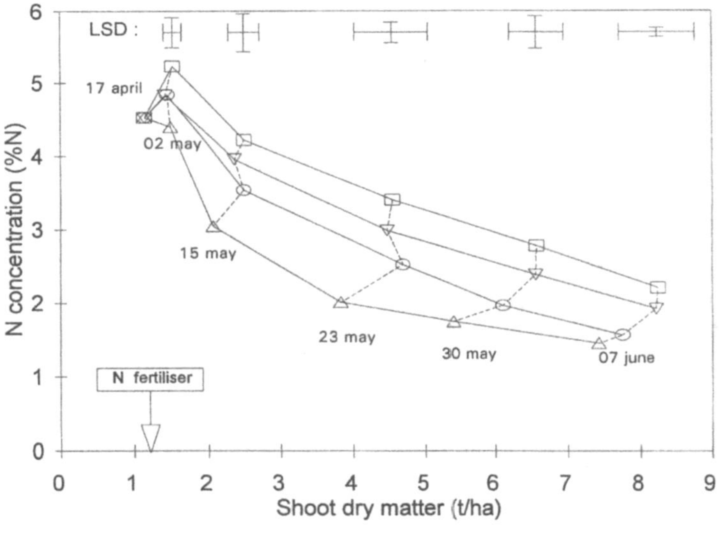

Fig. 4.1. Examples of four experimental plant N% curves. (Justes et al. 1994). The four treatments received 80 kg N ha^{-1} at tillering (Feb. 25, 1985) and the remainder at the stage ear at 1 cm (April 18, 1985). Remainder for N80 (△) = 0 kg N ha^{-1}; for N123 (○) = 43 kg N ha^{-1}; for N166 (▽) = 86 kg N ha^{-1}; and for N210 (□) = 130 kg N ha^{-1}. *LSD* Least significant difference

concentration of the aerial biomass decreases while growth rate remains constant; secondly, the growth rate decreases and dry matter accumulation is reduced.

4.2.1
Calculation of "Critical" Points

Definition. Justes et al. (1994) proposed that the N concentration of each point of the critical dilution curve should be defined as the minimum N concentration observed at a given date among all N treatments that had produced the maximum amount of dry matter up to that date. This corresponds to the N concentration at which the aerial dry matter does not significantly increase with an increase in N supply, despite an increase in crop N uptake and shoot N accumulation.

Application of Statistical Analysis. At each measurement date and for each experiment, the amounts of aerial dry matter (W) produced and the N concentrations (N%) were compared for different N fertilizer levels. The Student's bilateral t test was used to test the hypothesis of the means being equal at the 10% level of probability. The 10% significance level was preferred over the 5% significance level which is classically used, in order to reduce the occurrence of type II error (Dagnélie 1975). The experimental treatments retained to compute the critical total N concentration were composed of three to five points (W_i, N_i) corresponding to different fertilizer-N levels. Two cases can be distinguished depending on the position of these n points. Each set of points must satisfy the following inequalities (in the statistical meaning):

First case: $W_1 < W_2 = \cdots = W_n$ (n = 3, 4 or 5)
$N_1 < N_2 \leq \cdots < N_n$.

Second case: $W_1 < W_2 < W_3 = \cdots = W_n$ (n = 4 or 5)
$N_1 < N_2 < N_3 \leq \cdots < N_n$.

For each measurement date, the variation of N concentration in relation to shoot dry matter was described as a bilinear relationship comprising:

1. An oblique line representing a joint increase in dry matter and N concentration: in the first case above, this line passes through points D_1 and D_2; in the second case, it is the regression line between points D_1, D_2, and D_3, minimizing the sum of square errors.
2. A vertical line corresponding to an increase in N concentration without increase in the aerial biomass: its abscissa is equal to the mean abscissa of points D_2 to D_n (first case) or D_3 to D_n (second case).

In both cases, the critical N concentration corresponds to the ordinate of the intersection of the oblique and vertical lines.

4.2.2
Establishment of a Critical Plant N% Curve for Winter Wheat Crops

Among the experiments studied, only a small proportion of sampling dates (17%) fulfilled the previously defined statistical criteria. Each of these situations allowed us to compute, using the method previously described, a critical N concentration (total and reduced) for a given value of aerial biomass.

For values of W less than $1\,\mathrm{t\,ha^{-1}}$, four critical points could be calculated (Fig. 4.2): critical total N concentration (N_c) varied between 4.41 and 5.23% for approximately the same dry matter (0.8 to $0.9\,\mathrm{t\,ha^{-1}}$). In fact, the estimation of a critical concentration is less precise in these situations because the slope of the oblique line used in the calculation is very steep.

For values of W greater than $1\,\mathrm{t\,ha^{-1}}$, the critical points were estimated much more precisely, since the slope of the oblique line was always less than at lower biomass (Fig. 4.2). We fitted these data to different mathematical functions using the nonlinear regression program NEWSNL (Huet et al. 1992). Power functions (proposed by Salette and Lemaire 1981; Greenwood et al. 1990) and hyperbolic functions (proposed by Caloin and Yu 1986; Greenwood et al. 1991) fit the experimental data well, yielding to the following equations:

$$N_c = 5.35 \cdot W^{-0.442} \tag{1}$$

$$N_c = 0.77 + \frac{14.12}{2.48 + W}. \tag{2}$$

The power function [Eq. (1)] characterized well the decrease in the critical plant N% of winter wheat (Fig. 4.2); it can be applied when shoot dry matter lies in the

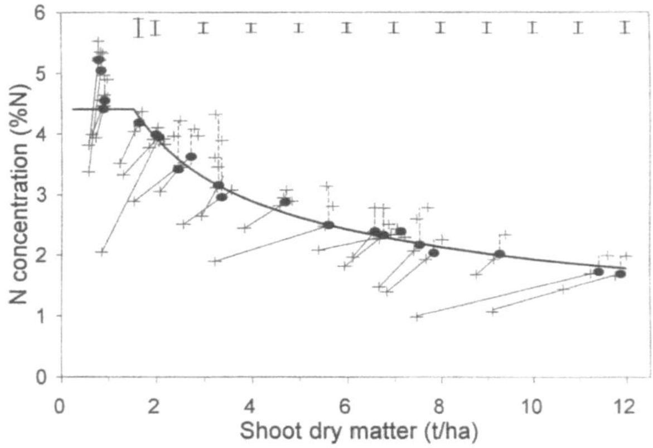

Fig. 4.2. Diagram representing N concentration versus accumulated dry matter in shoots, for selected data satisfying statistical criteria as defined in the text. (Justes et al. 1994). (+) Values measured for the different fertilizer N levels (D_1–D_5) at a given date; (●) "critical" points (N_c) calculated at each date (method shown in Fig. 4.1); (—) critical curve equation: $N_c = 5.35 \cdot W^{-0.442}$; *vertical bars* indicate confidence intervals on fitted critical curve ($P > 0.95$)

range 1.55–12 t ha^{-1} and for stages of development ranging from 1-cm-ear (Feekes 5) to the onset of flowering (Feekes 10.5.1).

Furthermore, if we consider a minimum N concentration (N_S) corresponding to the structural plant component (see Chap. 1), the power function becomes:

$$N_c - N_s = a \cdot W^{-b}. \tag{3}$$

However, this fit was not satisfactory, being slightly biased for biomass > 8 t ha^{-1}; thus we rejected this equation.

In addition, the critical dilution curve for reduced N was calculated as followed:

$$N_{cr} = 5.29 \cdot W^{-0.443}. \tag{4}$$

The critical curves of total N [Eq. (1)] and reduced N [Eq. (4)] are, in fact, very close, since the data points in the vicinity of the critical curve correspond to plants containing small quantities of N-NO_3^-: on average, 1% of the total plant N of these plants was present in the nitrate form. Gastal et al. (1992) also observed that fescue plants which had received an optimal N fertilizer application, and approached the critical N% curve, contained low quantities of N-NO_3^- (0.1%).

4.2.3
Validation of the Critical Curve for Winter Wheat Crops

The situations which were not retained under the statistical criteria defined earlier were used to test the validity of the critical curve. Thus, we carried out a new selection on the data in order to identify the significantly N-limiting and nonlimiting treatments. A treatment was considered to be N-limiting when its shoot dry matter W was significantly less ($P > 0.90$) than the shoot dry matter of the treatment that received the next highest rate of fertilizer N. Several treatments were considered N-nonlimiting when their shoot W values did not differ significantly ($P > 0.90$) although they had received different N rates.

Figure 4.3 shows that the critical curve previously derived was validated for all these situations within the biomass range used to establish it (1.65 to 12 t ha^{-1}). It was also validated for biomasses varying from 12 to 14 t ha^{-1} at flowering (Feekes 10.5.1 to 10.5.4). These results indicate that growth rate, density (between 150 and 400 plants m^{-2}), cultivar, and pedoclimatic conditions had little effect on the critical plant N%.

In contrast to the above, the equation of the critical curve cannot be applied to low biomasses (close to 1 t ha^{-1}). In the range 0.2–1.55 t ha^{-1} (Feekes 3 to 5), Justes et al. (1994) suggested that the critical concentration may be constant, and that a mean value of 4.4% could be used. Above 1.55 t ha^{-1}, the curve N_c [Eq. (2)] was applied corresponding to a value of W equal to 1.55 t ha^{-1}.

The range of N concentrations reported by different authors for optimal conditions of N nutrition is large: 4.8 to 5.5% (Jonard and Odent 1967), 3.5 to 4.5% (Karlen and Whitney 1980), 3.6 to 4.7% (Masle 1980). The value of 4.4% retained

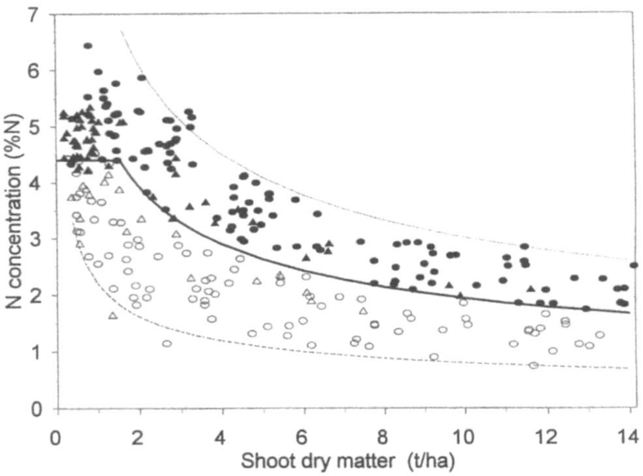

Fig. 4.3. Diagram representing N concentration versus accumulated dry matter in shoots, for all experimental treatments, N either limiting or nonlimiting for growth (Justes et al. 1994). *Open symbols* ○, △ Treatments where N was statistically limiting for shoot growth; *closed symbols* ●, ▲ treatments where N was statistically nonlimiting for shoot growth; *circles* ○, ● N determined with the Dumas method; *triangles* △, ▲ N determined with the Kjeldahl method; (—) critical plant N% curve [Eq. (3)]; (----) envelope curve N_{min} [Eq. (8)]; (......) envelope curve N_{max} [Eq. (9)]

is close to the critical concentration of 4.5% suggested by Meynard (1985) for wheat at the tillering stage, and to the optimal concentration of 4.29% proposed by Ingestad (1979) for leaves of seedlings of birch (*Betula pendula*) grown in controlled environment conditions. It equals the values of 4.4% measured by Jarvis (1987) for perennial ryegrass and by Devienne (1993) for spring wheat, if we consider only the reduced nitrogen content.

For Eq. (1), the decrease in critical N concentration in the shoots occurs only when W exceeds $1.55\,t\,ha^{-1}$, i.e., when the leaf area index (LAI) is around 2.5. According to Puckridge and Donald (1967) and Meynard (1985), competition for light among plants has already started at these values of LAI.

4.2.4
Validation of the Critical Curve for Other Cereal Crops

We performed a selection on data on malting barley (*Hordeum distichum* L.) sown in spring or autumn and durum wheat (*Triticum turgidum* ssp. *turgidum* conv. *durum*) from an experiment carried out by ITCF[4] in order to identify and retain the significantly N-limiting and nonlimiting treatments, as previously done for winter wheat (*Triticum aestivum* ssp. *vulgare*). As we obtained a large variability in some

[4] ITCF: Institut Technique des Céréales et des Fourrages (French Technical Institute of Cereal and Forage crops)

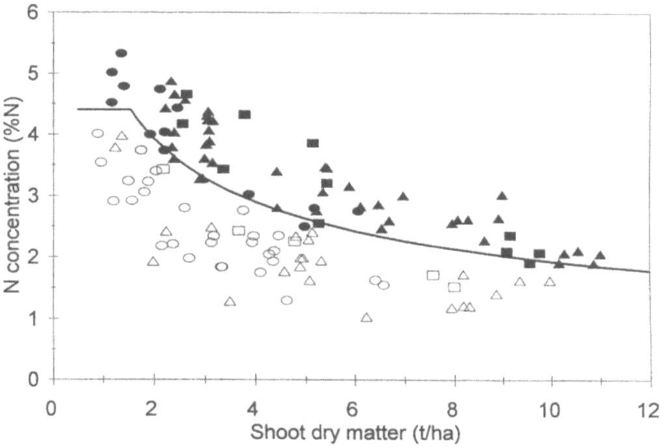

Fig. 4.4. Diagram representing total N concentration versus accumulated dry matter in shoots, for experimental treatments with N either limiting or nonlimiting for growth in spring barley and durum wheat (experiments from ITCF). *Open symbols* ○, △ Treatments where N was statistically limiting for shoot growth; *closed symbols* ●, ▲ treatments where N was statistically nonlimiting for shoot growth; *circles* ○, ● spring barley (*Hordeum distichum* L.); *triangles* △, ▲ durum wheat (*Triticum turgidum* ssp. *turgidum* conv. *durum*); (—) critical plant N% curve established for winter wheat (*Triticum aestivum* ssp. *vulgare*)

trials of durum wheat (coefficient of variation from 10 to 30%), we chose a probability level of $P < 0.15$ for durum wheat, slightly higher than that used for wheat and barley ($P < 0.10$).

Figure 4.4 shows the winter wheat critical plant N% curve and data for durum wheat and barley. Except for three nonlimiting data points that were slightly lower than the critical curve, the curve was validated for all the data sets for malting barley and durum wheat that were retained in the analysis over the biomass range used to establish it, that is 1 to 12 t ha^{-1}; this corresponded to Feekes stages 5 to 10 for malting barley and Feekes 6 to 10 for durum wheat. Thus we can consider that the critical plant N% curve established previously for winter wheat can be used for other cereals.

4.3 Link with Theory for Cereal Crops

We have shown that the critical curve for cereal crops can be defined by either a power function [Eq. (1)] or a hyperbolic function [Eq. (2)], at least for values of aerial dry matter ranging from 1.5 to 14 t ha^{-1}. Both functions fitted the observed data equally well and are consistent with the two theories explaining the nitrogen decrease of plant N% with increasing crop biomass (see Chap. 1).

It has been shown that the power function is possibly related to crop morphology, particularly to the surface: volume ratio of the crop. However, the power coefficient which has been found for the critical curve is equal to –0.44. This value is significantly lower than the –0.34 value proposed by Lemaire and Salette (1984) for grasses, and also different from the –0.33 value corresponding to the theoretical slope between ln(surface:volume) and ln(volume) of a simple solid such as a cylinder. However, this latter constant holds only for simple solid which expand proportionally in all directions. It is therefore not surprising to observe a different value for wheat crops, whose geometry is obviously more complex and whose relative expansion in the three dimensions may vary with ontogeny.

The hyperbolic function is possibly related to the decrease in the proportion of metabolic weight to the total weight (metabolic + structural) of the crop. According to one of the formulae proposed by Greenwood et al. (1991), Eq. (2) can be written as:

$$N_c = N_s + (N_m - N_s)\frac{x}{x+W}. \tag{5}$$

Identification of the terms allows us to calculate N_S and N_m: N_S = 0.77%; N_m = 6.47%. Unfortunately, these values cannot be determined very precisely, since they are correlated to each other in the fitting procedure. Statistical analysis indicates, for example, that the pair of values (0.70; 6.31) and (0.60; 6.11) are also possible for N_S and N_m. The value N_S = 0.70% was already proposed by Greenwood et al. (1991). The N concentration of the metabolic pool, N_m, should be close to the N concentration of a young seedling, whose proportion of structural material can be considered negligible. Surprisingly little information is available in the literature about this. F. Devienne (pers. comm.) found that wheat seedlings cultivated in sand with a nutrient solution containing a range of 5 to 20 mol N m^{-3} had a constant N concentration at the two leaves stage of 6.34 ± 0.02%. This result suggests that the 6.47% value for N_m may be realistic.

If we adopt the model of Caloin and Yu (1982) for describing growth in a canopy with no interplant shading [see Chap. 1, Eq. (19)], then the critical dilution curve could be described by the following two functions:

$$N_c = N_s + (N_m - N_s)\left(\frac{W_i}{W}\right)^a \quad \text{when } W_i \le W \le W_1 \tag{6a}$$

$$N_c = N_s + (N_m - N_s)\frac{x}{x+W} \quad \text{when } W \ge W_1 \tag{6b}$$

with the possible following constants:

$W_i = 0.004\,t\,ha^{-1}$	$W_1 = 1.5\,t\,ha^{-1}$	$x = 2.48\,t\,ha^{-1}$
$a = -0.08$	$N_s = 0.77\%$	$N_m = 6.47\%$.

This critical curve is represented in Fig. 4.5a (solid line). Since the proportion of metabolic to total dry matter in shoots is, by definition, given by:

$$\frac{W_m}{W} = \frac{N - N_s}{N_m - N_s}. \tag{7}$$

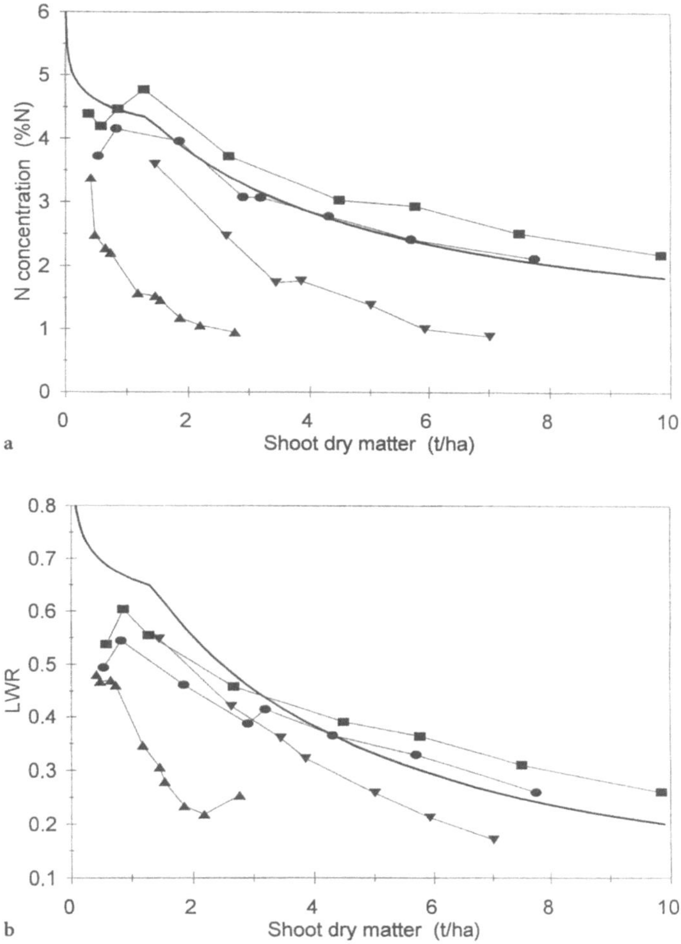

Fig. 4.5. a Time course of N concentration of wheat crops in four experimental treatments: ■ S_0 = no N stress (N fertilizer added regularly); ● S_1 = slight N stress (early deficiency); ▼ S_2 = large N stress (fertilizer added once only); ▲ S_3 = very severe N stress (no fertilizer added); (—) critical plant N% curve. **b** Time course of the ratio LWR of wheat crops in four experimental treatment (legend as in **a**)

The corresponding function is drawn in Fig. 4.5b (solid line).

These figures show the very rapid decrease in N which must occur in the very early stages of growth due to the rapid decrease in the relative size of the metabolic pool. The results are consistent with the relative stability of the critical N% for low shoot dry matter values that was already mentioned, at least for shoot biomass in the range 0.5 to 1.5 $t\,ha^{-1}$. However, the very rapid drop in N concentration during early growth remains to be confirmed in further studies. The hypothesis of metabolic and structural pools can also be tested by measurements of plant components. The theoretical ratio W_m/W can be compared to the ratio between the dry matter of green leaves and the total dry matter of the shoots, that is the leaf weight ratio (LWR). This was done in an experiment carried out in 1995 by M.H. Jeuffroy et al. (unpubl.) on wheat crops with different intensities of N deficiency. We present here four treatments denoted S_0 (no stress, due to regular N dressings), S_1 (slight stress), S_2 (moderate stress), and S_3 (very early and marked N stress). The time course of the N concentration and the LWR in the four treatments are given in Fig. 4.5a,b, respectively. A very close similarity is apparent in the time course of these two measures in each of the four treatments. The reduction in the proportion of green leaves is particularly obvious when N is strongly deficient.

Equation (6) enables us to calculate the theoretical W_m/W ratio for all treatments using the N_S and N_m values previously cited. This ratio can then be compared to the LWR, as illustrated in Fig. 4.6. It appears that a good correlation exists between the two indices ($r^2 = 0.81$, 50 df), but the regression line of W_m/W ratio on LWR has a slope (0.60) significantly different from 1, and an intercept (0.19) different from 0. This indicates that the theoretical W_m/W ratio cannot be determined by the proportion of green leaves to shoots alone. The two criteria are more or less equal for the treatments without N stress (S_0 and S_1), but differ widely for the strongly N-deficient treatments (S_2 and S_3).

We have verified that the discrepancy cannot be explained by the lack of precision in the determination of N_S and N_m values or by accounting for a fixed proportion of structural material in leaves and a fixed proportion of metabolic material in shoots. Therefore, this discrepancy confirms that the consequence of an N deficiency is not only a decrease in W_m/W ratio, but also a decrease in the N concentration in the metabolic compartment (see Chap. 1).

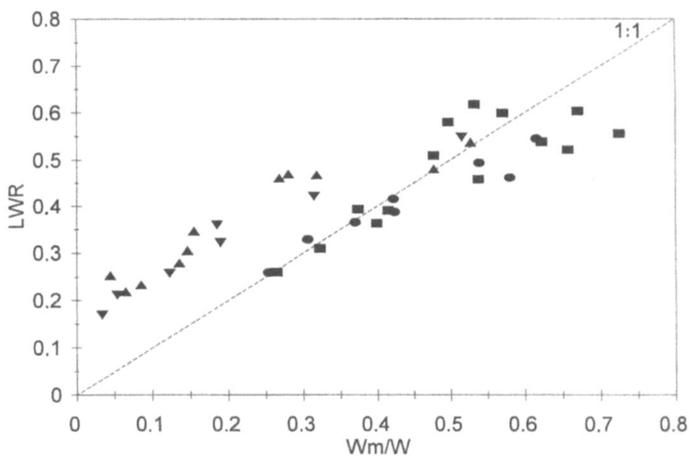

Fig. 4.6. Relationship between the LWR and the ratio W_m/W for the treatments shown in Fig. 5a,b. (----) Curve for LWR = ratio W_m/W

4.4 Use of Nitrogen Nutrition Index (NNI) to Analyze Yield Determination

4.4.1 Variability of N Uptake Versus Dry Matter Produced

Magnitude. We have observed a large variability in shoot N concentrations for a given value of shoot W in these experiments, although more variable environmental conditions can be found in wheat crops in France (see Fig. 4.3). We propose to define two envelope curves: a curve of minimum N concentration (N_{min}) valid from 0.5 to $14\,t\,ha^{-1}$, and a curve of maximum N concentration (N_{max}) valid from 1.5 to $14\,t\,ha^{-1}$. The envelope curves can be satisfactorily described by the following equations:

$$N_{min} = 2.1 \cdot W^{-0.44} \tag{8}$$

$$N_{max} = 8.5 \cdot W^{-0.44}. \tag{9}$$

For a given value of aerial biomass, the N concentration in the shoots can therefore vary by a factor of 4. It is not surprising to observe that total N concentrations can be 60% lower than the critical values in cases of severe N deficiency. It is interesting to note the magnitude of excess N consumption which can occur in crops grown on soils with a high N availability, the maximum N% curve being almost 60% above the critical dilution curve.

Proportion of Nitrate and Reduced Nitrogen. The large accumulation of N which may occur in shoots above the critical N% values (and therefore in conditions of maximum growth) has often been described as luxury consumption (Ågren and Ingestad 1987; Hirose et al. 1988; Millard 1988), and the extra N (over N_c) is often supposed to be nonassimilated N, present in the plant as nitrate.

Justes et al. (1994) showed that the fraction of nitrate N in shoots increased markedly as NNI increased, but found a large variability at a given level of N nutrition: the nitrate N varied between 0 and 18% of total-N. After stem elongation, it did not represent more than 8% of total N. It is also interesting to consider the ratio $f = N_n(N_t - N_c)$, i.e., the fraction of nitrate N to the extra N accumulated in shoots above the critical concentration, when N supply is not limiting aerial growth. This proportion of nitrate f was relatively high at tillering (about 50%) but much smaller later on (mean value: 20%), even for very high N contents. Therefore, most (80%) of the nitrogen accumulated above the critical curve was reduced N. These results are in agreement with those reported by Zhen and Leigh (1990) in controlled environment conditions.

N Absorption and Growth Control. The N_{max} curve [Eq. (9)] can be considered as a first estimate of the potential N uptake under field conditions (i.e., the maximum accumulation capacity of the shoots). It can be obtained from treatments where both growth and N accumulation rates are maximum. This curve is very different from the critical N% curve.

For plants which follow the maximum curve, the N absorption rate would be regulated by mechanisms directly associated with growth (Clement et al. 1978; Deane-Drummond 1984; MacDuff and Wild 1988; Gastal and Saugier 1989; Oscarson et al. 1989; Rufty et al. 1989; Raper et al. 1991) or indirectly associated with growth via N metabolism (Lee and Ratcliffe 1986; Cooper and Clarkson 1989; Touraine et al. 1992). By contrast, Angus and Moncur (1985) considered that growth is reduced as soon as shoot N decreases below N_{max}; this conclusion is in complete disagreement with our results obtained under field conditions.

Below the maximum curve, N absorption would be determined by the N availability in the soil. Between the critical curve and the maximum curve, N absorption would be regulated independently of growth rate, since the latter is maximal. Below the critical N% curve, both growth rate and N absorption would be determined by the availability of soil inorganic N. According to Penning de Vries (1982), the N_{min} curve [Eq. (8)] can be defined as the lower limit of N% at which the plant metabolism would cease to function.

4.4.2
Relationship Between the Nitrogen Nutrition Index (NNI) and Yield Formation

4.4.2.1
Nitrogen Deficiency and Yield Formation

In this section, we analyze the effects of temporary N deficiency and large values of N absorption (extra N over N_c) on seed number determination in a winter wheat crop, using the NNI [see Chap. 1, Eq. (21)] as an indicator of the deficiency. Since NNI can only be estimated until anthesis (see Sect. 4.2), and seed number is determined at this stage, the analysis is carried out for seed number only, and not for yield. Moreover, mean weight per grain depends mainly on crop photosynthesis between anthesis and maturity, and little on nitrogen nutrition.

We compare the grain number index (GNI) to NNI on a large range of wheat crops varying in N fertilization rates. The GNI is the ratio of grain number in each treatment to grain number in the treatment without N deficiency (NNI $\geq$ 1 at anthesis) in the same experiment, which corresponds to the optimum grain number. The range of data sets used corresponded to 7 years of experiments in the field (1984, 1985, 1989, and 1991–1994), 7 cultivars (Soissons, Thésée, Forby, Fidel, Appolo, Talent, Monopole), periods of N deficiency varying in dates of appearance and durations (due to various dates and rates of N application), and different regions of France varying in soil and climate. When there was an N deficiency (NNI < 1), the NNI measured at anthesis appeared to be significantly related to the grain number index (Fig. 4.7). In all these situations, the seed number was always lower than the seed number produced without N deficiency. The relationship between the two variables seems independent of cultivars. It is curvilinear: the lower the NNI, the greater the ratio GNI : NNI. NNI measured at anthesis can then be used to analyze the seed number produced by a wheat crop, or, more precisely, to estimate

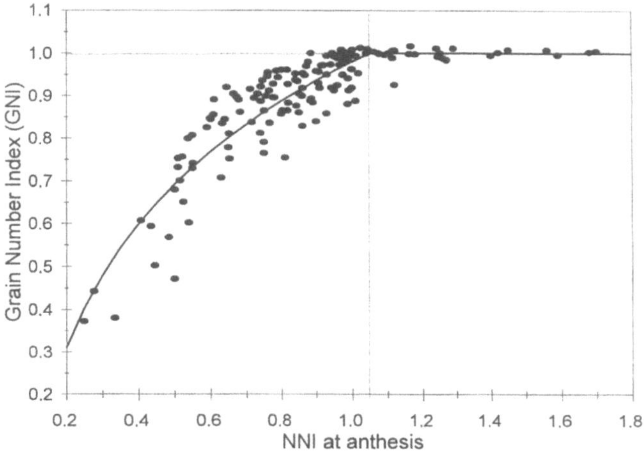

Fig. 4.7. Relationship between nitrogen nutrition index (NNI) at anthesis and grain number index (GNI), for different cultivars (including Soissons and Thésée) in various situations in France. Grain number index (GNI) = grain number in selected treatment : grain number in treatment without N deficiency in the same experiment.
Model: when NNI ≤ 1, GNI $= 0.98 + 0.42 * \ln(\text{NNI})$ $r^2 = 0.804$, 134 df,
when NNI > 1, GNI $= 1$

the deviation from the seed number which could have been produced if no nitrogen deficiency had occurred in the crop. According to the curve shown in Fig. 4.7, it seems that the maximum seed number can be reached only if NNI measured at anthesis is slightly greater than 1 (about 1.05); these results are in good agreement with those observed by Machet (1989). For NNI less than 1, the residual variability of GNI is rather high. It can be assumed that this residual variability depends partly on the date of occurrence of the N deficiency.

4.4.2.2
Relationship Between the Time Course of Change in NNI and Yield Formation

The effect of nitrogen nutrition during the various steps of seed formation along the crop cycle can be analyzed using the nitrogen index. This is illustrated in two experimental treatments presenting a very different nitrogen nutrition pattern between the beginning of stem elongation and anthesis of a wheat crop (Fig. 4.8). In each treatment, a period of N deficiency (NNI < 0.9) occurred for a period of 3 weeks, but at a different stage: from Feekes 5 (March 25) to Feekes 7 (April 17) in the ES (early stress) treatment, and from Feekes 7 (April 23) to Feekes 10 (May 14) in the LS (late stress) treatment. Fertilizer N was applied at the end of each stress period (150 kg ha^{-1} in ES, 50 kg ha^{-1} in LS), leading to the same amount of N accumulated in shoot at anthesis in the two treatments (135 kg ha^{-1}). The NNI in ES was very low at the beginning of the observation period and increased slowly after

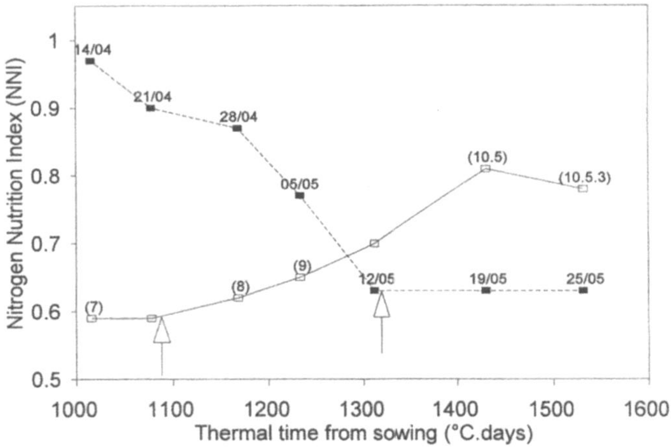

Fig. 4.8. Time course of nitrogen nutrition index (NNI), between 14/04 and anthesis (25/05), in two experimental treatments (ES and LS) varying in their periods of N deficiency, Grignon (France), 1992. □ ES treatment; ■ LS treatment; *arrows* indicates fertilizer N application; (*number*) developmental stages (Feekes' scale); *21/04* = Julian day

Table 4.1. Growth and yield components of winter wheat crop at various dates between the beginning of N deficiency and anthesis for two treatments, ES and LS (see text)

Treatment	Date	Stage	NNI	Stem (no./m^2)	LAI (m^2/m^2)	RUE (g/MJ)	W (g/m^2)	Grain (no./m^2)	Grain/Wa (no./g)
ES	21/04	2 Nodes	0.59	986	2.47	2.31[a]	243		
	25/05	Anthesis	0.78	453	3.33	2.54[a]	936	21 670	23.2
LS	21/04	2 Nodes	0.90	1045	3.84	2.96[a]	348		
	25/05	Anthesis	0.63	620	5.10	2.12[a]	1117	21 320	19.1

RUE = radiation use efficiency; LAI = leaf area index; W = dry matter; Wa = dry matter at anthesis.
[a] RUE was calculated between 21/04 to 05/05 for 1st stage and between 12/05 to 25/05 for 2nd stage.

the N application. In the LS treatment, NNI was close to 1 at Feekes 7 and declined slowly until the N application, near earing.

Table 4.1, which summarizes the growth and yield components, shows that at anthesis (May 25) the NNI reached significantly different values in the two treatments, whereas the grain number per unit area was very close. It is thus necessary to analyze the different steps of grain number formation to explain this discrepancy.

During stem elongation, the number of stems which stopped their growth and development was large in ES: this corresponded to a lower NNI at the beginning of stem elongation. This result is consistent with the model described by Masle (1985): in the case of N deficiency, only the tillers which have at least three leaves elongate and produce an ear.

The smaller number of stems in ES was one of the causes of the lower LAI observed in this treatment. However, the early N deficiency attested by the low NNI had another effect: it reduced the leaf area per stem. The direct effect of N limitation on the elongation rate of leaves and their final area was shown by Greenwood (1966) and Triboï and Ntonga (1993).

The radiation use efficiency (RUE) increased with time in ES, whereas it decreased in LS until anthesis. The variations in RUE were correlated with the variations in NNI. This is consistent with the fact that leaf photosynthetic capacity depends on leaf nitrogen content (e.g., Hunt 1985; review of Sinclair and Horie 1989). A correlation between RUE and NNI was already found by Belanger et al. (1992) for tall fescue.

The smaller amount of radiation intercepted after the beginning of stem elongation, linked to the lower LAI, caused a smaller biomass production in ES (cf. models from Monteith 1972, 1977). In spite of the increase in RUE during the earing and flowering stages (second period of RUE calculation) linked to the increase in NNI, the dry matter produced at anthesis was still smaller in ES than in LS. Thus, the accumulation of dry matter in ES was reduced very early on, and it never reached the level of LS during the cycle.

The ratio between grain number and dry matter at anthesis (N efficiency) was greater in ES than in LS. It is known that the seed number produced is highly dependent on the availability of assimilates (Fischer and Stockman 1980; Brooking and Kirby 1981; Stockman et al. 1983), particularly during the month preceding anthesis (Fischer 1985), but the N deficiency in LS during this sensitive period (attested by the low NNI) could also have induced abortion of grain, as observed by Darwinkel (1983) and Triboï and Ntonga (1993). This would explain the absence of significant difference in seed number between the two treatments.

4.4.2.3
Relationship Between NNI and Grain Protein Content

In all the situations shown previously in Fig. 4.7, the correlation between NNI measured at anthesis and grain protein content at harvest was low. This is generally the case where N absorption after anthesis is large. When we exclude the cases where N accumulated in shoots after anthesis comprised more than 55% of the total amount of N taken up by maturity, the correlation is significant. The value of NNI at anthesis is also a valuable indicator of the grain protein content for a given wheat cultivar (Fig. 4.9). This result is consistent with the fact that N accumulated in the grains is mainly derived from N absorbed before anthesis (Jonard and Odent 1967; Novoa and Loomis 1981; Herzog 1986). The residual variability of the grain protein content, which is rather large, is linked to various adverse conditions that may occur during grain filling (high temperature, water stress, etc.). In these situations, the accumulation of dry matter in the grain is reduced, accompanied by a relatively large protein content. A significant relationship was also observed in barley, but with NNI measured at earing (Gede Wibawa 1992).

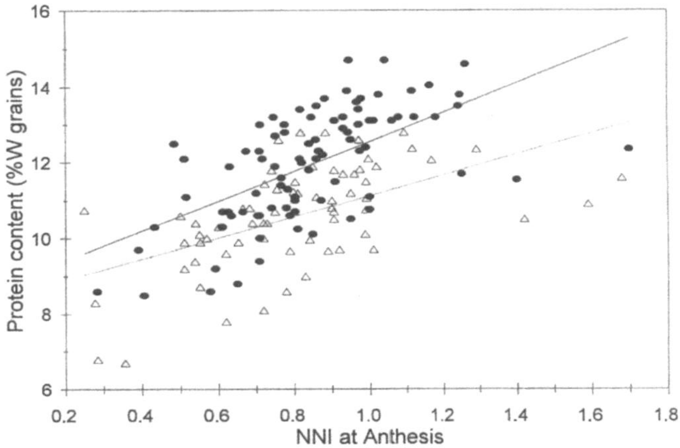

Fig. 4.9. Relationship between grain protein content (N%. 5.7) and nitrogen nutriton index (NNI) at anthesis, for various situations in France, for cultivars Soissons and Thésée. ● = Cultivar Soissons; (—) = linear regression for Soissons (r^2 = 0.36, 86df); △ cultivar Thésée; (......) = linear regression for Thésée (r^2 = 0.31, 67df)

4.5 Conclusion

The results indicate the validity of the critical curve of plant N%, for different conditions of growth and development, cultivars and population densities in wheat, barley, and durum wheat. It has been validated from tillering to flowering, and enables a precise N nutrition diagnosis during the growth cycle to flowering.

The nitrogen nutrition index (NNI) derived from this curve is a synthetic and integrative indicator of plant nitrogen status. The change in the time course of NNI is a good indicator of the yield determination before anthesis of cereal crops. The results presented in this chapter indicate that NNI can be used to characterize the different effects of N deficiency on the successive steps of seed number formation. In crop simulation models presented in the literature, the effect of N deficiency is assumed to be mainly on growth, due to reduction in photosynthesis (van Keulen and Seligman 1987; Groot and de Willigen 1991; Hansen et al. 1991), and sometimes variations in stem and leaf number (Porter 1993). These models often use the curve of maximum N accumulation in the plant to define the occurrence of N deficiency (Penning de Vries 1982; Angus and Moncur 1985; Porter 1993), and not the critical N% as defined above. Clear reference parameters to quantify the N stress are often missing in the present models as outlined by Van Keulen and Stol (1991). The use of the critical curve and the NNI could help to improve them.

Acknowledgments. We wish to thank François Laurent (ITCF), who authorized the use of data of some wheat experiments and all the data for spring barley and durum wheat.

References

Ågren GI, Ingestad T (1987) Root:shoot ratio as a balance between nitrogen productivity and photosynthesis. Plant Cell Environ 10:579–586

Angus JF, Moncur MW (1985) Models of growth and development of wheat in relation to plant nitrogen. Aust J Agric Res 36:537–544

Belanger G, Gastal F, Lemaire G (1992) Growth analysis of a tall fescue sward fertilized with different rates of nitrogen. Crop Sci 32:1371–1376

Brooking IR, Kirby EJM (1981) Interrelationships between stem and ear development in winter wheat: the effects of a Norin 10 dwarfing gene, Gai/Rht_2. J Agric Sci Camb 97:373–381

Caloin M, Yu O (1982) An extension of the logistic model of plant growth. Ann Bot 49:599–607

Caloin M, Yu O (1986) Relationship between nitrogen dilution and growth-kinetics in Graminae. Agronomie 6:167–174

Clement CR, Hopper MJ, Jones LHP (1978) The uptake of nitrate by *Lolium perenne* from flowing nutrient solution. J Exp Bot 29:453–464

Cooper HD, Clarkson DT (1989) Cycling of amino-nitrogen and other nutrients between shoots and roots in cereals. A possible mechanism integrating shoot and root in the regulation of nutrient uptake. J Exp Bot 40:753–762

Dagnélie P (1975) Théories et méthodes statistiques. Applications agronomiques. II. Les méthodes de l'interférence statistique. Les presses Agronomiques de Gembloux, Gembloux

Darwinkel A (1983) Ear formation and grain yield of winter wheat as affected by time of nitrogen supply. Neth J Agric Sci 31:211–225

Deane-Drummond CE (1984) Mechanism of nitrate uptake into *Chlara corallina* cells: lack of evidence for obligatory coupling proton pump and a new NO_3^-/NO_3^- exchange model. Plant Cell Environ 7:317–323

Devienne F (1993) Influence de la concentration en nitrate du milieu sur les flux d'absorption d'azote et la croissance du blé. Thèse de Doctorat, INA Paris-Grignon

Fischer RA (1985) Number of kernels in wheat crops and the influence of solar radiation and temperature. J Agric Sci Camb 105:447–461

Fischer RA, Stockman YM (1980) Kernel number per spike in wheat (*Triticum aestivum* L.): responses to preanthesis shading. Aust J Plant Physiol 7:169–180

Gastal F, Saugier B (1989) Relationships between nitrogen uptake and carbon assimilation in whole plants of tall fescue. Plant Cell Environ 12:407–418

Gastal F, Belanger G, Lemaire G (1992) A model of the leaf extension rate of tall fescue in response to nitrogen and temperature. Ann Bot 70:437–442

Gede Wibawa (1992) Approche par enquête et expérimentation de l'effet de l'état structural du sol sur la nutrition azotée et l'élaboration du rendement de l'orge de brasserie. Thèse de Doctorat, INA Paris-Grignon

Greenwood DJ, Lemaire G, Gosse G, Cruz P, Draycott A, Neeteson JJ (1990) Decline in percentage N of C_3 and C_4 crops with increasing plant mass. Ann Bot 66:425–436

Greenwood DJ, Gastal F, Lemaire G, Draycott A, Millard P, Neeteson JJ (1991) Growth rate and N% of field grown crops: theory and experiments. Ann Bot 67:181–190

Greenwood EAN (1966) Nitrogen stress in wheat. Its measurement and relation to leaf nitrogen. Plant Soil 24:279–288

Groot JJR, de Willigen P (1991) Simulation of the nitrogen balance in the soil and a winter wheat crop. Fertil Res 27:261–272

Hansen S, Jensen HE, Nielsen NE, Svendsen H (1991) Simulation of nitrogen dynamics and biomass production in winter wheat using the Danish simulation model DAISY. Fertil Res 27:245–259

Herzog H (1986) Source and sink during reproductive period of wheat. Advances in Agronomy and Crop Science. Parey, Berlin

Hirose T, Freijsen AHJ, Lambers H (1988) Modelling of the responses to nitrogen availability of two *Plantago* species grown at a range of exponential nutrient addition rates. Plant Cell Environ 11:827–834

Huet S, Jolivet E, Méssean A (1992) La régression non linéaire. Méthodes et applications biologiques. INRA Editions, Paris

Hunt LA (1985) Relationships between photosynthesis, transpiration and nitrogen in the flag and penultimate leaves of wheat. In: Day W, Atkin RK (eds) Wheat growth and modelling. NATO ASI Series. Plenum Press, New York, pp 149–156

Ingestad T (1979) Nitrogen stress in birch seedlings. II. N K P Ca and Mg nutrition. Physiol Plant 45:149–157

Jarvis SC (1987) The effects of low, regulated supplies of nitrate and ammonium nitrogen on the growth and composition of perennial ryegrass. Plant Soil 100:99–112

Jonard P, Odent M (1967) Etude sur l'évolution de l'azote au cours de la croissance de la tige principale du blé tendre. Ann Amélior Plant (Paris) 17:23–31

Justes E, Mary B, Meynard J-M, Machet J-M, Thelier-Huché L (1994) Determination of a critical nitrogen dilution curve for winter wheat crops. Ann Bot 74:397–407

Karlen DL, Whitney DA (1980) Dry matter accumulation, mineral concentrations, and nutrient distribution in winter wheat. Agron J 72:281–288

Large EC (1954) Growth stages in cereals. Illustration of the Feekes scale. Plant Pathol 3:129

Lee RB, Ratcliffe RG (1986) Effects of nitrogen deficiency on the absorption of nitrate and ammonium by barley plants. Ann Bot 57:471–486

Lemaire G, Salette J (1984) Relation entre dynamique de croissance et dynamique de prélèvement d'azote pour un peuplement de graminées fourragères. I. Etude de l'effect du milieu. Agronomie 4:423–430

Macduff JH, Wild A (1988) Changes in NO_3^- and K^+ uptake by four species in flowing solution culture in response to increased irradiance. Plant Physiol 74:251–256

Machet JM (1989) Relationships between N nutrition and the yield elaboration of winter wheat. In: Ecophysiologie du blé, Les Colloques de l'INRA, Versailles

Masle J (1980) L'élaboration du nombre d'épis chez le blé d'hiver. Influence de différentes caractéristiques de la structure du peuplement sur l'utilisation de l'azote et de la lumière. Thèse Docteur-Ingénieur, INA-PG, Paris

Masle J (1985) Competition among tillers in winter wheat: consequences for growth and development of the crop. In: Day W, Atkin RK (eds) Wheat growth and modelling. NATO ASI Series. Plenum Press, New York, pp 33–54

Meynard JM (1985) Les besoins en azote du blé d'hiver jusqu'au début de la montaison. Agronomie 5:579–589

Millard P (1988) The accumulation and storage of nitrogen supply by herbaceous plants. Plant Cell Environ 11:1–8

Monteith JL (1972) Solar radiation and productivity in tropical ecosystems. J Appl Ecol 9:747–766

Monteith JL (1977) Climate and the efficiency of crop production in Britain. Philos Trans R Soc Lond B 281:277–294

Novoa R, Loomis RS (1981) Nitrogen and plant production. Plant Soil 58:177–240

Oscarson P, Ingemarson B, Larsson CM (1989) Growth and nitrate uptake properties of plants grown at different relative rates of nitrogen supply. II. Activity and affinity of the nitrate uptake system in *Pisum* and *Lemna* in relation to nitrogen availability and nitrogen demand. Plant Cell Environ 12:779–785

Penning de Vries FWT (1982) Crop production in relation to availability of nitrogen. In: Penning de Vries FWT, van Laar HH (eds) Simulation of plant growth and crop production. Simulation monographs. PUDOC, Wageningen, pp 213–233

Porter JR (1993) AFRCWHEAT2: a model of the growth and development of wheat incorporating responses to water and nitrogen. Eur J Agron 2:69–82

Puckridge J, Donald S (1967) Competition among wheat plants sown at a wide range of densities. Aust J Agric Res 18:193–211

Raper CD Jr, Vessey JK, Henry LT (1991) Increase in nitrate uptake by soybean plants during interruption of the dark period with low intensity light. Bot Gaz 139:289–294

Rufty TW, MacKown CT, Volk RJ (1989) Effects of altered carbohydrates availability on whole-plant assimilation of $^{15}NO_3^-$. Plant Physiol 89:457–463

Salette J, Lemaire G (1981) Sur la variation de la teneur en azote des graminées fourragères pendant la croissance: Formulation d'une loi de dilution. CR Acad Sci Paris Sér III 292:875–878
Sinclair TR, Horie T (1989) Leaf nitrogen, photosynthesis, and crop radiation use efficiency: a review. Crop Sci 29:90–98
Stockman YM, Fischer RA, Brittain EG (1983) Assimilate supply and floret development within the spike of wheat (*Triticum aestivum* L.). Aust J Plant Physiol 10:585–594
Touraine B, Muller B, Grignon C (1992) Effect of phloem-translocated malate on NO_3^- uptake by roots of intact soybean plants. Plant Physiol 99:1118–1123
Triboï E, Ntonga J (1993) Effet de l'azote et du rayonnement sur le développement des feuilles et de l'épi chez le blé d'hiver: mise en place de l'appareil foliaire et de la structure de l'épi. Agronomie 13:253–265
van Keulen H, Seligman NG (1987) Simulation of water use, nitrogen nutrition and growth of a spring wheat crop. Simulation monographs. PUDOC, Wageningen
van Keulen H, Stol W (1991) Quantitative aspects of nitrogen nutrition in crops. Fertil Res 27:151–160
Zhen RG, Leigh RA (1990) Nitrate accumulation by wheat (*Triticum aestivum*) in relation to growth and tissue N concentrations. Plant Soil 124:157–160

CHAPTER 5

Maize and Sorghum

D. Plénet[1] and P. Cruz[2,3]

Contents

5.1 Introduction

The optimization of fertilizer N application to maize crops is fundamental to avoid pollution of ground water by nitrate. In maize, the great variability of N soil mineralization in summer, the importance of animal manure inputs in many situations, and the uncertainty of the potential production as a consequence of variation in the intensity of summer drought, make prediction of the optimum N application rate very difficult.

Until now the calculation of the recommended rate of N application has been based on the estimated soil N balance. This empirical method allows the estimation of (1) soil N mineralization for the crop growth period and (2) the crop N demand necessary to achieve expected potential yield. Measurement of soil mineral N at the beginning of the growth period can improve the accuracy of the predictions (Magdoff et al. 1992). The N demand of the crop is generally calculated as the product of the target yield by N requirement per yield unit (biomass or grain). This determination is very imprecise, because of frequent differences between the target yield and the actual yield as a result of unforeseeable weather

[1] INRA Unité d'Agronomie, 71, av. Edouard Bourleaux, BP 81, 33883 Villenave d'Ornon Cédex, France
[2] INRA Unité Agropédoclimatique des Caraïbes, BP 515, 97165 Pointe-à-Pitre Cédex, France
[3] Present address: INRA Station d'Agronomie, BP27, 31326 Castanet-Tolosan Cédex, France

Diagnosis of the Nitrogen Status in Crops
G. Lemaire (ed)

conditions. Thus, this method often leads to a large discrepancy between N supply and crop N demand.

Some diagnostic procedures have been proposed in order to verify the adequacy of N supply to N requirement of a crop: foliar diagnosis (Loué 1984) or the DRIS system (Walworth and Sumner 1987). However, these techniques are based on a determination of N concentration in a particular organ of the plant (ear leaf in maize, for instance), and therefore depend strongly on phenological stages. These methods do not account for the dynamics of plant growth.

5.2 The Decline in Critical Plant N% in Maize and Sorghum Crops

Maize and sorghum crops are able to take up greater quantity of N than is necessary for their maximum growth. Thus the determination of crop N requirement implies determination of the critical N uptake, i.e., the minimum N uptake to achieve maximum crop biomass. As presented in Chapter 1, critical N uptake determination corresponds to determination of the critical plant N%.

5.2.1 Critical Plant N% for Maize

An experimental network was set up during 1990–1994 in the southwest of France with irrigated maize crops (cv. Volga, Pioneer, FAO index = 550) with different N application rates (three to six rates according to the year). Crop biomass and plant N% were recorded from successive samplings. Among the 58 data points we obtained 26 situations where the critical plant N% could be determined according to the statistical method of Justes et al. (1994). The data are presented in Fig. 5.1 using a log/log scale. The decline in critical plant N% can be described by two regression lines, the first from the ten-leaf stage to the 23rd day after flowering with a slope of −0.37, and the second from the 23rd day after flowering with a steeper slope. This acceleration of the decline in the plant N% corresponds to the end of vegetative growth of the plants and enhancement of the grain filling process. Figure 5.2 presents the decline in the critical plant N% ($\%N_c$) with crop mass (W) for the period of vegetative growth:

$$\%N_c = 3.40(W)^{-0.37}. \tag{1}$$

This relationship applies to crop mass in the range 1–22.0 t ha^{-1}. For values of crop mass less than 1 t ha^{-1} a constant value of $\%N_c$ can be adopted, as proposed in Chapter 1. The value of 3.4% which is used is close to the optimal value for maize lamina determined by Loué (1984) and Plénet (1995) in field conditions, and also corresponds to the optimum leaf N content for photosynthesis (Khamis and Lamaze 1990).

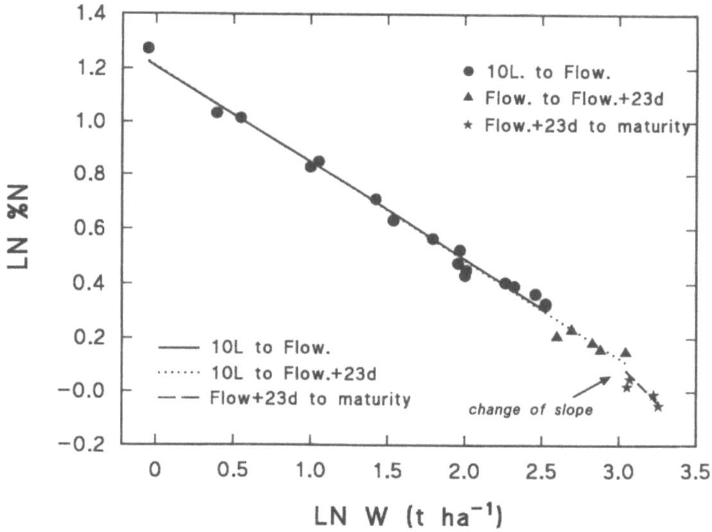

Fig. 5.1. Relationship between the log of aerial biomass (*W*) and the log of critical N concentration in the dry matter of the aerial biomass (*%N*), for different stages of maize crops

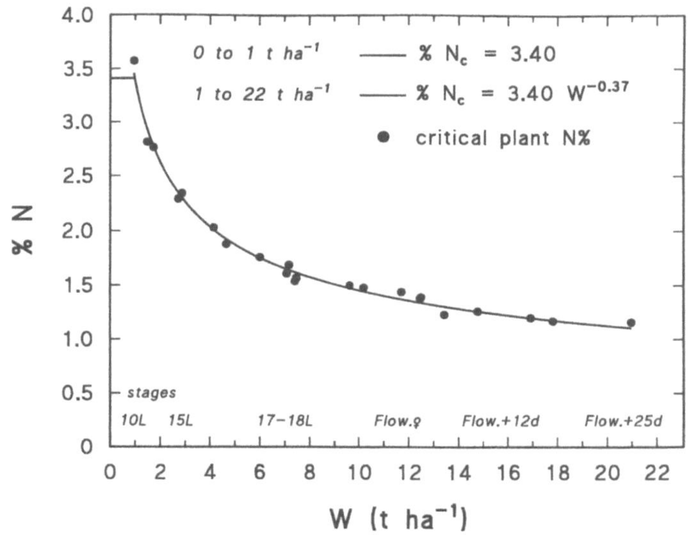

Fig. 5.2. Critical plant N% versus aerial biomass (W) for development stages from ten leaves to flowering +23 days. Critical N curve obtained by nonlinear fitting in the range 1–22 t DM ha^{-1}. For W < 1 t DM ha^{-1}, a constant value of 3.4% is used

This relationship between critical plant N% and crop mass has been validated using a data base from a large experimental network using the same variety in other pedoclimatic regions. The data points were classified in two categories: (1) nonlimiting N conditions, when no statistical difference in crop mass (Student test with $P < 10\%$) was observed between different levels of N application; in this case, all the corresponding data points were considered as nonlimiting; and (2) limiting N conditions, when statistical differences in crop mass were detected. In this case, only the lower levels were considered as limiting. Figure 5.3 shows that the critical plant N%-crop mass relationship discriminates well between the limiting and the nonlimiting situations. Similar studies carried out with other genotypes of different maturity dates (cv. Furio and cv. Dea) confirmed the accuracy of Eq. (1).

Figure 5.3 allows the calculation of two curves which envelop all the data points:

- Minimal envelope curve: $\%N_{min} = 2.05(W)^{-0.56}$.
- Maximal envelope curve: $\%N_{max} = 6.30(W)^{-0.42}$.

For a given crop mass, the nitrogen nutrition index (NNI), defined in Chapter 1 as the ratio actual plant N%:critical plant N%, can vary from 1.8 to 0.3. The minimal envelope curve cannot be considered as the absolute minimum of plant N%, especially for the early stages of growth when soil N, even if restricted, cannot be totally nil. If we draw a horizontal line on Fig. 5.3 corresponding to a minimum plant N%, the estimated value of 0.7 seems close to the value of 0.8 proposed in

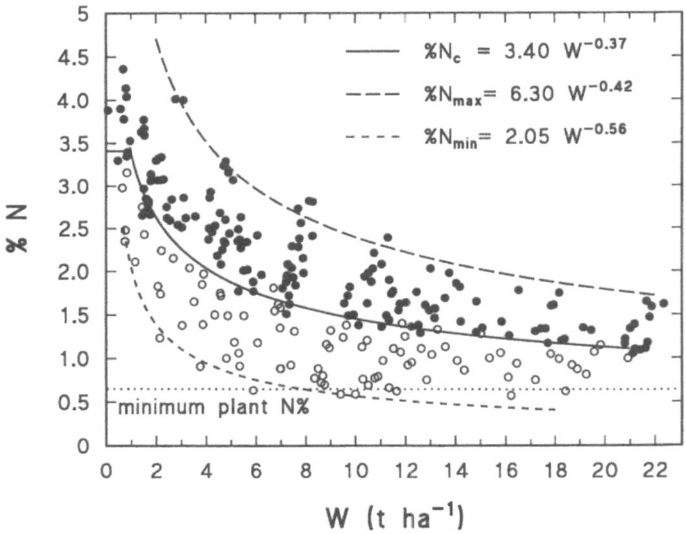

Fig. 5.3. Relationships between %N and aerial biomass (W) for different experiments. • Treatments where N was statistically limiting for shoot growth; ○ treatments where N was not statistically limiting for shoot growth; — critical curve: $\%N_c = 3.40\,W^{-0.37}$; --- envelope curve $\%N_{max}$ and $\%N_{min}$; .... estimation of a constant minimum plant N%

Chapter 1 for structural plant N%. The maximal envelope curve gives an estimation of the maximum capacity of N accumulation by the maize crop. Clearly, the maize crop is able to accumulate a large amount of excess N. In terms of diagnosis for N fertilization management, it is important to take this possibility into account and to analyze to what extent plants are able to use this excess N to prevent (or delay) N deficiency during subsequent periods when soil N availability becomes limiting.

From Eq. (1), it is possible to determine the relationship between critical N uptake and crop mass:

$$N_c = 34(W)^{0.63}. \tag{2}$$

with N_c in kg N ha^{-1} and W in t ha^{-1}; 34 kg N ha^{-1} corresponding to the quantity of N necessary to produce a crop biomass of 1 t ha^{-1}; and 0.63 corresponding to the allometric coefficient between N uptake and crop growth rate.

This relationship applies from a crop biomass of 1 t ha^{-1} (around the stage of ten visible leaves) to a stage corresponding to about 600 degree-days after flowering (about 25–30 days after flowering, with a crop mass of about 22–23 t ha^{-1}). At this stage about 95% of the total N which is absorbed by the crop at harvest has already been taken up. After this time, further N absorption from the soil is very limited, and grain filling is mainly supplied by remobilization of N from leaves and stems.

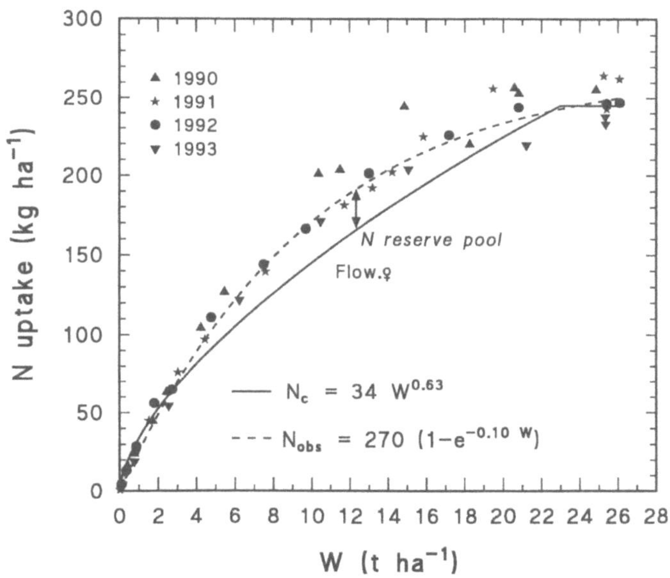

Fig. 5.4. Relationship between nitrogen uptake (kg N ha^{-1}) and the accumulation of aerial biomass (W in t ha^{-1}). — N_c curve: critical crop N demand; --- N_{osb} curve: minimal envelope curve of the N uptake observed on the nonlimiting N treatments in the experiments carried out in 1990–1993

Equation (2) allows calculation of the critical crop N demand for the maximum biomass accumulation rate by the maize crop; but as shown in Fig. 5.3, this relationship does not represent the maximum N uptake profile of the crop, which is more related to the N soil availability and to the rate and the date of N application. Figure 5.4 shows that to allow the crop to reach its critical N uptake at the end of the growth period, the application of N fertilizer in the early period leads to uptake of an excess of N which can be considered as an N reserve pool. In most of the situations tested, the soil N availability at the end of the growing period was generally insufficient to supply crop N demand, and temporary "luxury" consumption allowed the crop to achieve its growth potential. Thus, the diagnosis of crop N nutrition status during the vegetative period must take into account the necessity of an N reserve pool of about 25 kg N ha^{-1}, because N applications late in season are not possible (except when N application can be associated with irrigation water).

5.2.2 Critical Plant N% for Sorghum

Experimental data on sorghum are less numerous than for maize, and it was not possible to use the same methodology for the determination of critical plant N% because only three N treatments were available for each experiment. The data base was created with the following rules:

- When no statistical differences were detected between the three N treatment ($N_1 = N_2 = N_3$), N_1 was considered critical and was introduced in the data base.
- When $N_1 < N_2 = N_3$, the N_2 treatment was introduced in the data base.
- When $N_1 \ll N_2 < N_3$, the N_3 treatment was considered as close to the critical point and was introduced in the data base.

This procedure could have led to a slight overestimation of the critical plant N% because the two first situations were more frequent, particularly during the beginning of the growth period. Figure 5.5 shows the relationship obtained for two types of sorghum: sudangrass cv. Piper, and Sudan grass x sorghum hybrid cv. Sweet Sioux. An acceleration of the rate of decline in plant N% occurred for the two genotypes just after the flowering stage. For the period of vegetative growth, a single relationship between plant N% and crop mass can be proposed for the two varieties:

$$\%N_c = 3.90(W)^{-0.39}. \qquad (3)$$

The critical plant N% for a crop mass of 1 t ha^{-1} appears to be higher for sorghum (3.9%) than for maize (3.4%). This difference is probably due to the overestimation of critical plant N% of sorghum for the early stage of growth. This overestimation leads to a slight underestimation of the regression slope: –0.39 vs. –0.37. Thus, despite the difficulties in determining exactly the decline in critical plant N% for the sorghum crop, we can conclude that the critical curves for the

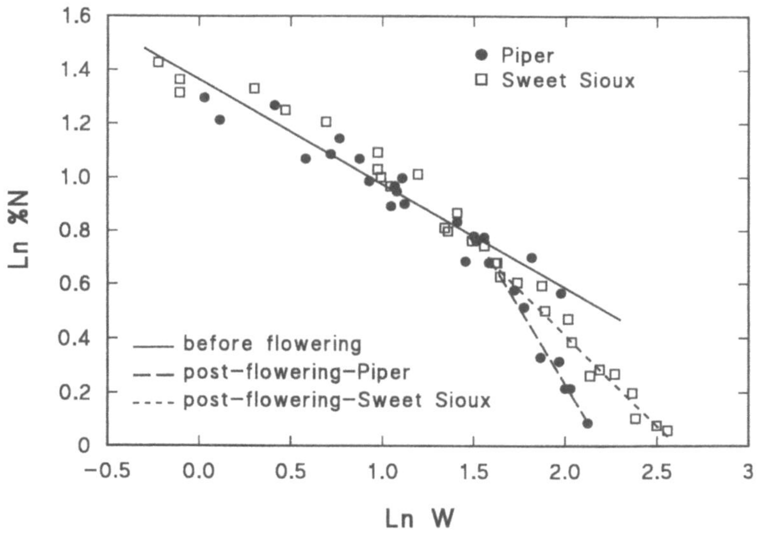

Fig. 5.5. Relationship between the log of aerial biomass (W) and the log of critical N concentration in the dry matter of the aerial biomass (%N), for different stages of sorghum crops. ● Sudan grass cv. Piper; □ Sudan grass x sorghum hybrid cv. Sweet Sioux

two crops are similar. A single relationship between critical N uptake and crop mass can therefore be proposed for both C_4 crops:

$$N_c = 34(W)^{0.63}. \quad (4)$$

This relationship indicates that the metabolic efficiency of nitrogen, i.e., the quantity of biomass produced per unit of N uptake, is identical for sorghum and maize.

5.3 Nitrogen Nutrition Index

The nitrogen nutrition index (NNI) can be calculated for the two crops by using the same critical plant N% curve according to the procedure explained in Chapter 1. Figure 5.6 shows the time course of the values of NNI observed in a maize crop on the same site during 2 successive years. Similar N application rates led to very contrasting values of NNI, indicating large variations in crop N status as a result of differences in soil N availability.

This example shows clearly the usefulness of the determination of NNI values for better interpretation of the experimental data. Depending on the rapid variations in soil N availability and the other soil conditions which interact with N absorption (e.g., soil water deficit), a given N application rate can lead to very

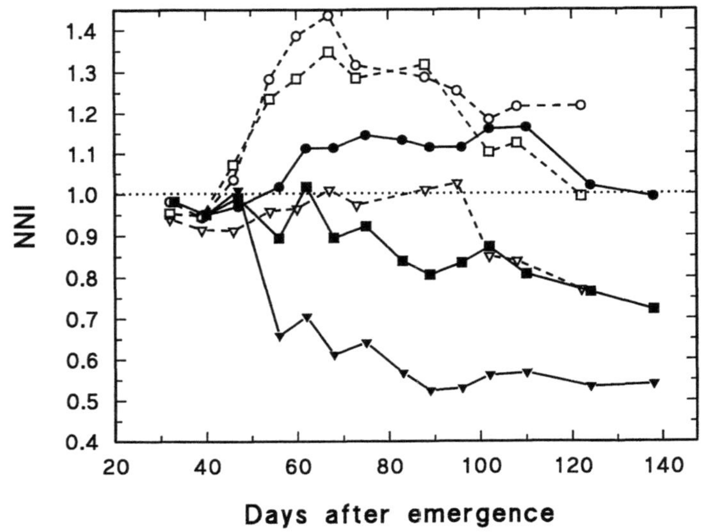

Fig. 5.6. Time course of the nitrogen nutrition index (NNI) for maize crops cultivated on the same site in 1990 and 1991. 1990 data: ○ 240 kg N ha^{-1}; □ 120 kg N ha^{-1}; ▽ 50 kg N ha^{-1}; 1991 data: ● 240 kg N ha^{-1}; ■ 100 kg N ha^{-1}; ▼ 30 kg N ha^{-1}

different N nutrition status in the crop. This makes it difficult to carry out on unbiased comparison of experimental results between years and sites. However, use of NNI makes such comparisons possible.

5.4 Nitrogen Nutrition Index and Productivity of Maize Crops

The use of the approach of Monteith (1977) allows the study of the effect of the nitrogen nutrition status of a crop on its productivity. The differences in production induced by the differences in N nutrition can be attributed to (1) the effect of N on the quantity of PAR absorbed by the crop, and (2) the effect of N on the radiation use efficiency (RUE). Plénet (1995) showed that the relative reduction of maize crop yield is clearly related to the value of NNI weighted as a function of time (Fig. 5.7).

The response of the LAI of the crop to N deficiency is similar to that of the aerial biomass, but the quantity of absorbed PAR is influenced relatively less by N shortage. This difference in the sensitivity of response to N between LAI and absorbed PAR is due to the fact that moderate N deficiencies occurred only after the optimum LAI was reached, and therefore had only a small effect on absorbed PAR. On the other hand, severe N deficiencies start earlier and therefore they greatly affect the amount of PAR absorbed by the crop. Thus, the overall effect of N deficiency on total PAR absorbed by the crop depends not only on the overall

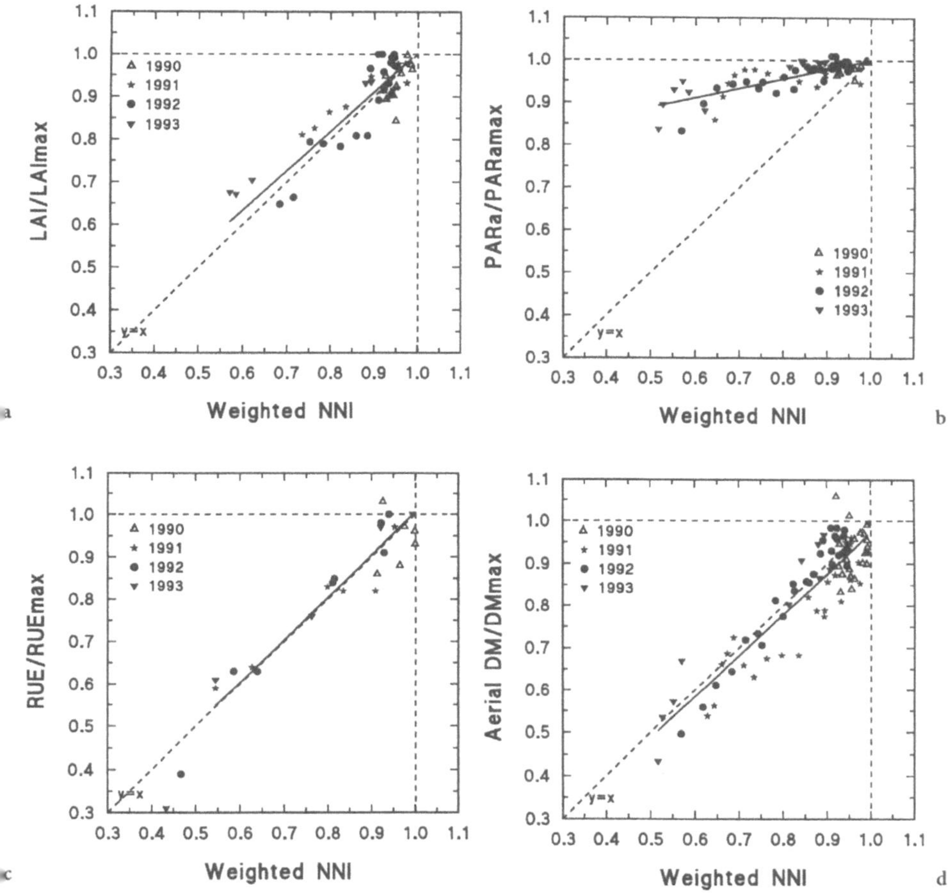

Fig. 5.7a–d. Relationships between nitrogen nutrition index weighted as a function of time (weighted NNI) and four relative variables of maize growth for 4-year experiments. The relative variables were calculated as the ratio between their values under limiting N conditions and their values under nonlimiting N conditions. **a** Relative leaf area index (LAI/LAI_{max}). **b** Relative cumulative absorbed photosynthetically active radiation ($aPAR/aPAR_{max}$). **c** Relative radiation-use efficiency (RUE/RUE_{max}). **d** Relative aerial dry matter (aerial DM/DM_{max})

intensity of the N shortage, but also on the relationship of the N deficiency period to the stage of leaf area development. RUE is strongly reduced by shortage of N. The relative decrease of RUE has the same magnitude as the decrease of NNI for values ranging between 0.55 and 1. For lower values of NNI, the RUE seems to be more severely reduced. This high sensitivity of RUE to N deficit was also observed by Muchow and Davis (1988) and Sinclair and Horie (1989) on sorghum and maize.

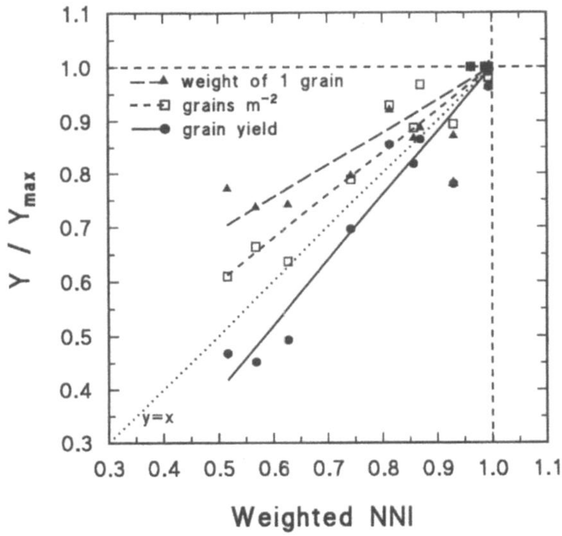

Fig. 5.8. Relationships between nitrogen nutrition index weighted as a function of time (weighted NNI) and relative yield components (Y/Y_{max}) of maize for 4-year experiments. The relative variables were calculated as the ratio between their values at limiting N conditions and their values at the nonlimiting N conditions. ▲ Relative weight per grain; □ relative number of grains m^{-2}; ● relative grain yield

Figure 5.8 allows an analysis of the effect of N deficiency on the yield components of the maize crop. The average grain weight is less sensitive to N deficiency than the number of grains per m^2, mainly for NNI lower than 0.7. This result confirms the data of Plénet et al. (1990b). On the other hand, the grain yield is more reduced than aerial biomass by N shortage.

The nitrogen nutrition index can be used at different growth stages to provide for the conditions of N nutrition necessary to obtain the maximum values of yield components. At the flowering stage of maize, that is 20 days before the date when the number of grains per unit area is set, maximum value of number of grains can be obtained only when NNI is greater than or equal to 1.0–1.05.

5.5 Comparison Between Maize and Sorghum

Figure 5.9 shows the relationship between the relative crop mass and the nitrogen nutrition index for sorghum and maize experiments. The data on maize were obtained during 4 years in France, while the data on sorghum come from experiments carried out over several years in France and Guadeloupe (French West Indies). In spite of imprecision in estimating the relative crop mass in such experiments, the two crops seem to respond similarly to N deficiency, and for sorghum it was not possible to discern differing responses according to the climatic condi-

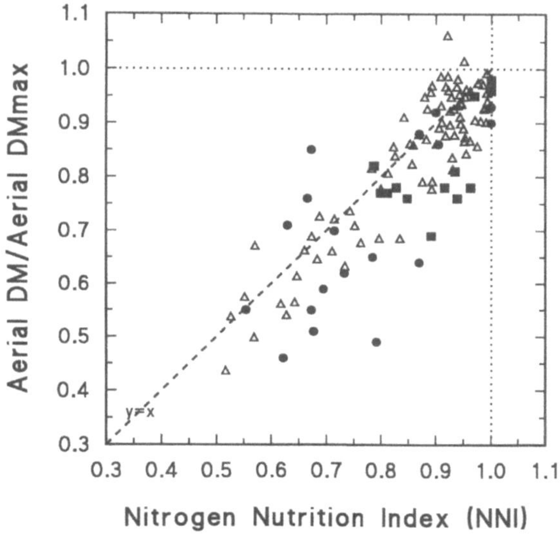

Fig. 5.9. Comparison between relative aerial biomass (aerial DM/DM_{max}) in maize and sorghum crops as a function of the nitrogen nutrition index (weighted NNI). △ Data on maize were shown Fig. 5.7d. Data on sorghum from ■ France and ● Guadeloupe

tions, temperate or tropical. Thus, we can conclude, as assumed in Eq. (4), that the metabolic efficiency of the two species is about the same with a similar N requirement per unit of crop biomass produced. Moreover, the crop biomass of the two species is similarly reduced with decrease in NNI.

This absence of a difference in N use efficiency between the two species does not imply that there is no difference in their capacity to capture soil N resources and to meet their N demand. Lemaire et al. (1996) compared the N uptake of maize and sorghum grown on the same soil under irrigated conditions with two N supply: 50 and 200 kg N ha^{-1}. The N status for the two crops was determined with NNI (results not presented). Figure 5.10 shows the N uptake as a function of aerial biomass in maize and sorghum. With a high rate of N application of 200 kg N ha^{-1}, the two crops accumulated the same quantity of N (280 kg ha^{-1}) at the end of the growth period; but due to the lesser growth of sorghum, this quantity of N represents to an excess of 70 kg ha^{-1} for this crop, while the same amount of N is necessary just to satisfy the N requirement of the maize. With a lower N application rate of 50 kg N ha^{-1}, the sorghum crop accumulated approximately the same quantity of N as with the higher N level, but with a delayed uptake kinetic, and therefore maintained its N nutrition at a nonlimiting level. With this lower N application rate the maize crop accumulated only 190 kg N ha^{-1}, which was insufficient to maintain its N nutrition at a nonlimiting level. In conclusion, the superiority of sorghum over maize for soil N extraction is due to (1) a lower growth rate which, by determining a lesser minimum daily N requirement, allows the crop to use a greater proportion of the available soil N before N deficiency occurs, and (2) a higher capacity for extraction of soil N in conditions of low availability in the soil. The latter property mainly depends on intrinsic characteristics of the root system, as suggested by Lemaire et al. (1996).

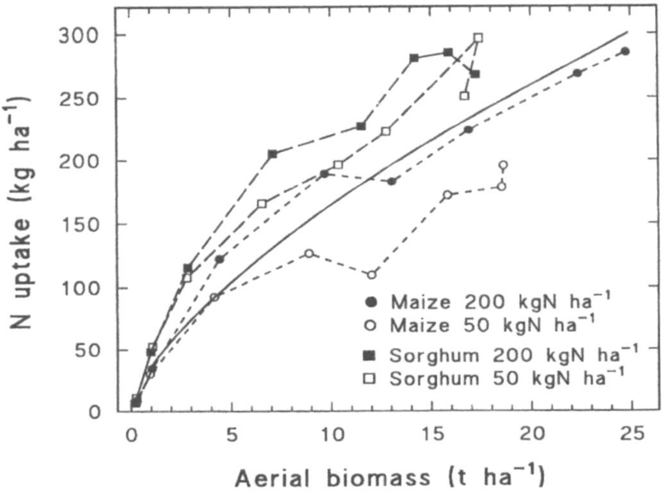

Fig. 5.10. Comparison of N uptake (kg N ha^{-1}) as a function of aerial biomass (t ha^{-1}) in maize and sorghum cultivated on the same soil under irrigated conditions with two N supplies: 50 and 200 kg N ha^{-1}. — Critical crop N demand: $N_c = 36\ (DM)^{0.66}$. (Lemaire and Chartier 1992)

5.6
Use of the Nitrogen Nutrition Index for N Fertilization Management

The technique of applying N through irrigation water tends to be developed in the southwest of France. This technique allows N applications after the 12-visible-leaf stage, which represents the latest stage for application of N in solid form. A synthesis of a network of 16 experiments carried out in France on this theme showed that, on average, dividing a given quantity of N into two or three applications gives no increase in grain yield compared to a single application at the beginning of growth, that is before the 6–8-visible-leaf stages (see, for instance, Plénet et al. 1990a). However, other authors showed that this technique could lead to significant increases in grain yield (Rehm and Wiese 1975; Rhoads and Stanley 1981). The effect of the late application could be dependent on the N status of the crop at this date. Unfortunately, it was not possible to obtain information on the NNI at the different dates of N application for these experiments. In a series of three new experiments, we observed that when the value of NNI was superior or equal to 1, the final N application was ineffective and could be omitted, but when the NNI was inferior to 1, the supplemental application always gave an increase in grain yield (Plénet 1995). Thus, the determination of the NNI value of maize crops at different growth stages allows a decision to be made whether or not to apply extra N according to the crop N status.

When a deficiency in crop N nutrition is detected, the effect of a corrective N application must be estimated in order to calculate the rate of N application required and to evaluate its benefit in terms of yield. Plénet (1995) found that maize crops which experienced values of NNI = 0.85 at the flowering stage showed a reduction in grain yield of less than 4% if a corrective N application was immediately applied, but if NNI reached a value of 0.55, the reduction in yield was 15–25%, whatever the corrective N application rate.

5.7 Conclusion

For maize, a relationship between critical plant N% and crop mass allows the determination of a nitrogen nutrition index, which can be used as a tool for diagnosis of the actual N nutrition status of a crop. Used concurrently with a crop growth model, this N diagnostic tool allows a quantitative interpretation of the discrepancy between actual and potential crop growth in terms of crop N status, whatever the origin of the N deficiency: insufficient N application rate or N shortage induced by drought. We have confirmed that the relationship between critical plant N% and crop mass is similar for the two crops, maize and sorghum, as postulated previously by Greenwood et al. (1991) for all C_4 species. The use of the nitrogen nutrition index has allowed a comparison of the N uptake capacity of the two species. Thus, we can confirm the greater capacity of sorghum compared to maize to extract N from soil and to satisfy the crop's N requirement when soil N availability is low.

References

Greenwood DJ, Gastal F, Lemaire G, Draycott A, Millard P, Neeteson JJ (1991) Growth rate and %N of field grown crops: theory and experiments. Ann Bot 67:181–190

Justes E, Mary B, Meynard JM, Machet JM, Thelier-Huche (1994) Determination of a critical nitrogen dilution curve for winter wheat crops. Ann Bot 74:397–407

Khamis S, Lamaze T (1990) Maximal biomass production can occur in corn (*Zea mays*) in the absence of NO_3 accumulation in either leaves or roots. Physiol Plant 78:388–394

Lemaire G, Chartier M (1992) Relationships between growth and dynamics of nitrogen uptake for individual sorghum plants growing at different plant densities. In: Scaife A (ed) Proc 2nd congr of European Society for Agronomy, 23–28 August 1992, Warwick, pp 98–99

Lemaire G, Charrier X, Hébert Y (1996) Nitrogen uptake capacities of maize and sorghum crops in different nitrogen and water supply conditions. Agronomie 16:231–246

Loué A (1984) Maïs. In: Martin-Prével P, Gagnard J, Gautier P (eds) L'analyse végétale dans le contrôle de l'alimentation des plantes tempérées et tropicales. Technique et documentation, Lavoisier, Paris, 810 pp

Magdoff FR, Jokela WE, Griffin GF, Klausner SD (1992) Predicting N fertilizer needs for corn in humid regions: advances in the Northeast. In: Bock BR, Kelleys KR (eds) Predicting N fertilizer needs for corn in humid regions. Bull Y-226. National fertilizer and environmental research center, Tennessee Valley Authority, Muscle Shoals, Alabama, 127 pp

Monteith JL (1977) Climate and the efficiency of crop production in Britain. Philos Trans R Soc Lond Ser B 281:277–294

Muchow RC, Davis R (1988) Effect of nitrogen supply on the comparative productivity of maize and sorghum in a semi-arid tropical environment. II. Radiation interception and biomass accumulation. Field Crops Res 18:17–30

Plénet D (1995) Fonctionnement des cultures de maïs sous contrainte azotée. Détermination et application d'un indice de nutrition. Thesis Docteur, Institut National Polytechnique de Lorraine, Nancy, 247 pp

Plénet D, Lubet E, Desvignes P, Girardin P (1990a) Effects of splitting of N application on N uptake and grain yield of irrigated maize crop. In: Scaife A (ed) Proc 1st Congr of European Society for Agronomy, Paris 5–7 Dec 1990, Session 1, p 34

Plénet D, Lubet E, Desvignes P, Sombrum F (1990b) Fertilisation azotée et composantes du rendement du maïs: effets des niveaux et des modalités d'apport. In: Picard D (ed) Colloq Physiologie et production du maïs, 13–15 Nov, Pau. INRA, Paris, 501 pp

Rehm GW, Wiese RA (1975) Effect of method of nitrogen application on corn (*Zea mays* L.) grown on irrigated sandy soils. Soil Sci Soc Am Proc 39:1217–1220

Rhoads FM, Stanley RL (1981) Fertilizer scheduling, yield and nutrient uptake of irrigated corn. Agron J 73: 971–974

Sinclair TR, Horie T (1989) Leaf nitrogen, photosynthesis and crop radiation use efficiency: a review. Crop Sci 29:90–98

Walworth JL, Sumner ME (1987) The diagnosis and recommendation integrated system (DRIS). Adv Soil Sci 6:149–188

CHAPTER 6

Grain Legumes

B. Ney[1], T. Dore[1], and M. Sagan[2]

Contents

6.1 Introduction

Previous chapters have described the general pattern of plant N% decrease with biomass accumulation in certain crops. In contrast to these crops, grain legumes have specific traits that could modify this pattern. Growth of the pods and of seeds with a high N concentration after flowering could compensate partly or totally for the decline in plant N%, caused by the decrease of the leaf: stem ratio and the remobilization of N from the shaded to the illuminated parts of the canopy. Theoretically, after the growth of pods and seeds, plant N% changes according to the number of reproductive organs that are growing. If seed number is high, nitrogen can be remobilized from roots, and shoot N% increases. Low seed number corresponds to a reduced shoot N%, as the capacity for N in the seeds is limited. Since the number of pods and seeds depends on many factors other than N nutrition, plant N% can decrease without N deficiency, simply because a large proportion of pods and seeds has aborted. Thus, the comparison between observed plant N% and critical plant N% could be inadequate to diagnose N nutrition deficiency after pod emergence.

[1] INRA-INA PG Laboratoire d'Agronomie, 78850 Thiverval-Grignon, France
[2] INRA Station de Génétique et Amélioration des Plantes, Domaine d'Epoisses, 21110 Bretenières, France

Diagnosis of the Nitrogen Status in Crops
G. Lemaire (ed)

Our aim in this chapter is to analyze plant N% decrease within grain legume crops and to discuss the value of the critical plant N% concept in managing grain legume crops. Several experiments have been conducted:

1. To compare the pattern of plant N% decrease in pea (*Pisum sativum* L.), soybean (*Glycine max* L.; Munier-Jolain 1994), lupin (*Lupinus albus* L.; Duthion and Pigeaire 1991), and lucerne (*Medicago sativa* L.; Lemaire and Allirand 1993) grown in contrasting environmental conditions.
2. To establish the critical plant N%:crop biomass relationship for this type of crop. A non-nodulating genotype of pea supplied with different regimes of nitrogen fertilizer as described by Sagan et al. (1993), was used for this purpose.
3. To test the performance of the nitrogen nutrition index (NNI) as defined above in evaluating the N nutrition of grain legume crops. The relationships between NNI at various stages and yield components have first been established experimentally and then used to analyze the influence of environmental conditions on the N status of pea crops in farm fields.

6.2 Plant N% in Grain Legume Crops in Relation to Developmental Stage

The available data collected for legume crops show conflicting results. As shown in Chapter 1 and by Lemaire and Allirand (1993), plant N% decreased with biomass accumulation in lucerne, as in cereals and other forage crops, but this was not the case in lupin canopies, at least between anthesis and the beginning of seed-filling, as shown by Duthion and Pigeaire (1991). These authors showed that plant N% remained constant during this period and was not influenced by the year, N fertilization regime, and cultivars within a given crop, for inoculated plants. Plant N% was found to vary only according to plant density and between N nutrition regimes for noninoculated plants. Soybean has also been found to behave in this way (Fig. 6.1a). Soybean was sown in 1991 in Dijon (France) at 18 and 60 plants m^{-2}, in Guadeloupe (French West Indies) at 50 plants m^{-2}, and in 1992 in Dijon at 20 plants m^{-2} (Munier-Jolain 1994). The results showed that, in tropical or temperate climates, plant N% remained almost constant during the growth period. Experiments carried out with pea sown in Dijon in March 1987, 1990, 1991, 1992, and 1994 at 80 plants m^{-2}, showed that the behavior of this crop was intermediate between that of lucerne and soybean (Fig. 6.1b). Until biomass reached approximately 1 t ha^{-1}, plant N% increased with biomass, then decreased and finally remained stable. These results suggest the important role of reproductive development in plant N% changes during plant growth. Until emergence of the pods, plant N% declines as a consequence of the increase in crop mass as described before for other crops. After pod emergence and during seed filling, the accumulation of N in reproductive organs with a high N concentration counterbalances the decline in N% of the vegetative plant components (leaves and stems), and therefore plant N% remains more or less constant.

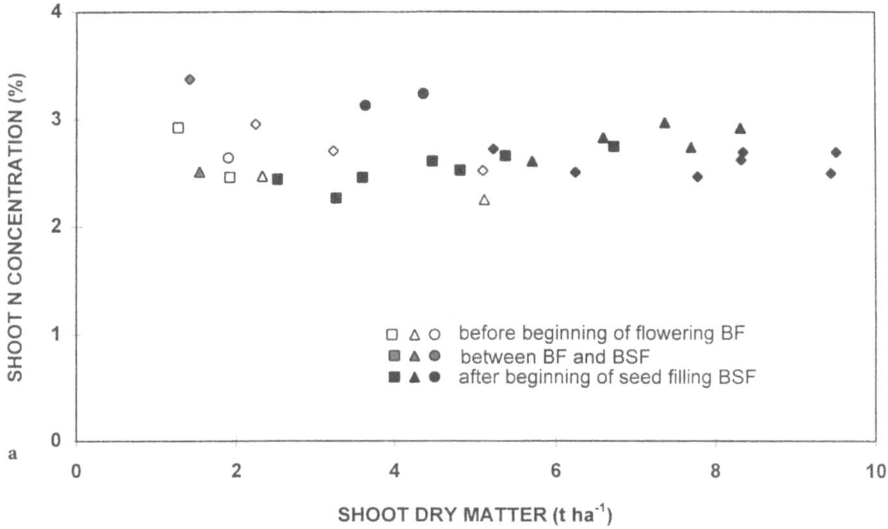

a

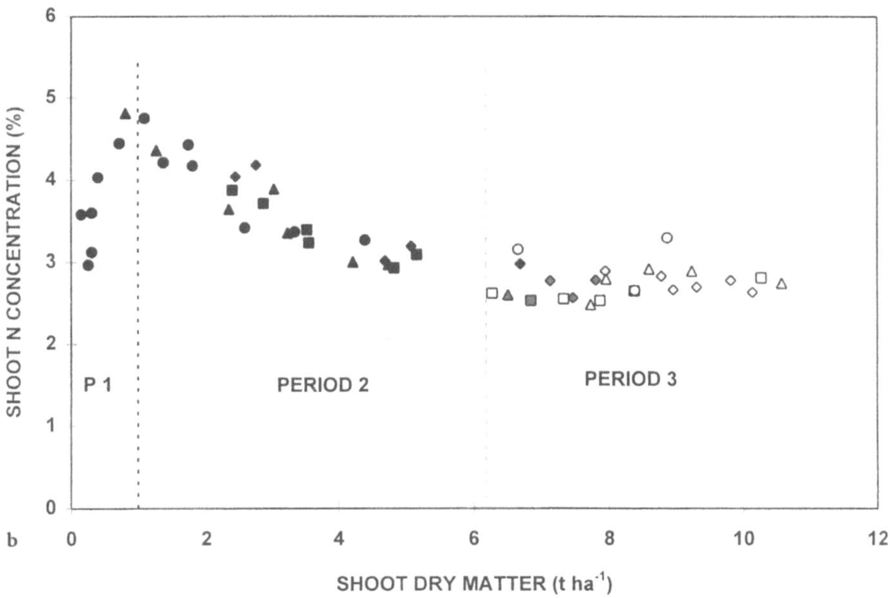

b

Fig. 6.1. **a** Shoot N concentration and biomass accumulation in soybean. **b** Shoot N concentration and biomass accumulation in pea cv. Frisson. **a** Dijon 1991: ■ 18 plants m^{-2}, ◆ 60 plants m^{-2}; Dijon 1992: ▲50 plants m^{-2}; Guadeloupe 1991: ●20 plants m^{-2}. **b** ■ 1987; ◆ 1991; ▲ 1992; ● 1994

The influence of the development stage on changes in the plant N% of legume crops could explain the apparent diversity of behavior among species. Lucerne, when cultivated as forage, does not produce seeds, and therefore its plant N% follows the general pattern described in Chapter 1. For soybean, in contrast, flowering and emergence of pods occur very early in France, when the biomass of the crop is near the limit of $1\,t\,ha^{-1}$. Therefore, the plant N% remains constant, as shown in Fig. 6.la. For pea, the situation is intermediate. Flowering occurs at greater crop biomass. After the beginning of seed-filling, which occurs at 250 degree-days (basic temperature 0 °C) after flowering (Ney and Turc 1993), plant N% decline is stopped by N accumulation in pods and seeds, as for soybean. Thus, depending on the earliness of the flowering stage, plant N% decrease in grain legume crops will cease at different values of plant N%.

Thus, despite these characteristics, critical plant N% approach appears to be adequate for legumes but only until the beginning of seed filling. During the seed-filling period, plant N% depends on the ratio between the seeds and other plant components. Late stresses that lead to pod and seed abortion, such as water stress, affect this ratio and hence influence plant N%. The method could be unreliable in this instance and is, in any case, inadequate for crops such as soybean, where seed-filling occurs early in growth. On the other hand, for pea, the period until the beginning of seed-filling is long, and the diagnosis with the critical plant N% concept can be reliable.

6.3 Determination of Critical Plant N% for Pea

Symbiotic mutants of pea used to define the critical plant N% curve (Sagan et al. 1993) were following methods used for cereals (Justes et al. 1994). Non-nodulating (P2) mutants and an isogenic fixating genotype, Frisson, were fertilized with different amounts of nitrogen at the time of sowing (0 or $25\,g\,N\,m^{-2}$ in 1990 and 1991, and 0, 12.5, and $25\,g\,N\,m^{-2}$ in 1993), or with $25\,g\,N\,m^{-2}$ at sowing date and $25\,g\,N\,m^{-2}$ at flowering in 1993. Plants were sown in the field at 60 to 80 plants m^{-2} in March at Dijon (France). Shoot dry matter per unit ground area and shoot N% (Kjeldahl) were recorded.

The critical plant N% curve was fitted to the points corresponding to the N fertilization rate above which plant N% increased without significant increase of dry matter (Fig. 6.2). The fitted equation was:

$$N\% = 5.08(DM)^{-0.32}, \qquad (1)$$

where N% and DM are shoot N% and shoot dry matter ($t\,ha^{-1}$). Coefficients of Eq. (1) are very close to those found for lucerne (Lemaire and Allirand 1993; Lemaire et al. 1985) and for C_3 species in general (see Chap. 1).

The curve described by Eq. (1) corresponds closely to the lower values of shoot N% at a given biomass found for the fixating genotype Frisson in experiments

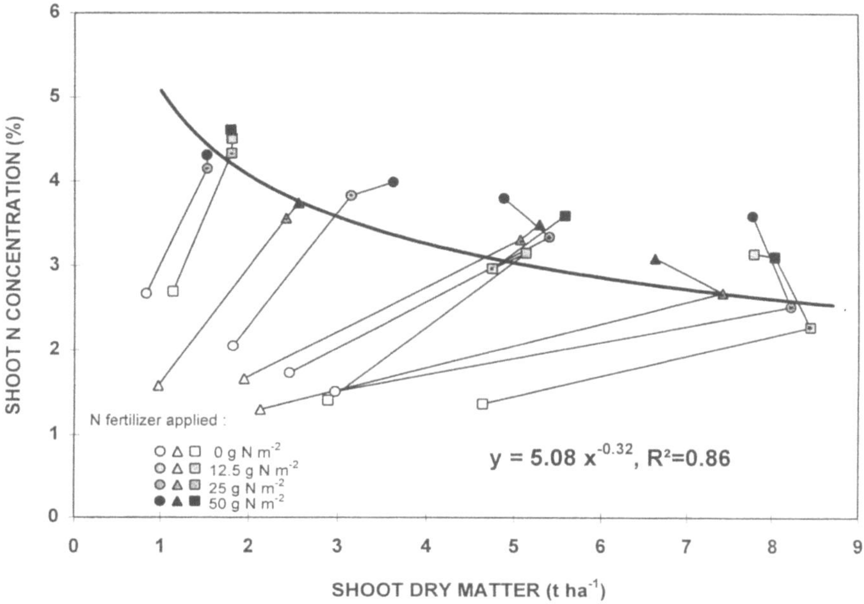

Fig. 6.2. Shoot N concentration and accumulation of biomass for non-nodulating mutants of pea cv. Frisson under different N regimes. ○ 1990; △ 1991; □ 1993; ⊙ ◬ ⊡ above this N fertilizer level, N concentration increases while dry matter does not increase significantly. N critical curve is fitted to these points

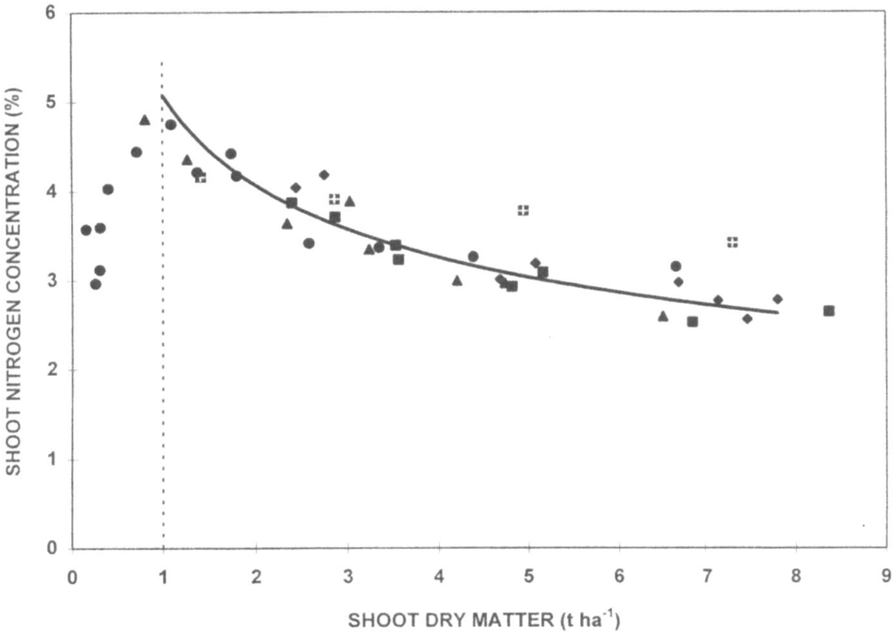

Fig. 6.3. Nitrogen dilution and critical curve Nc until the beginning of seed filling for pea cv. Frisson. ■ 1987; ● 1990; ◆ 1991; ▲ 1992; ● 1994; ___ N Critical curve Nc

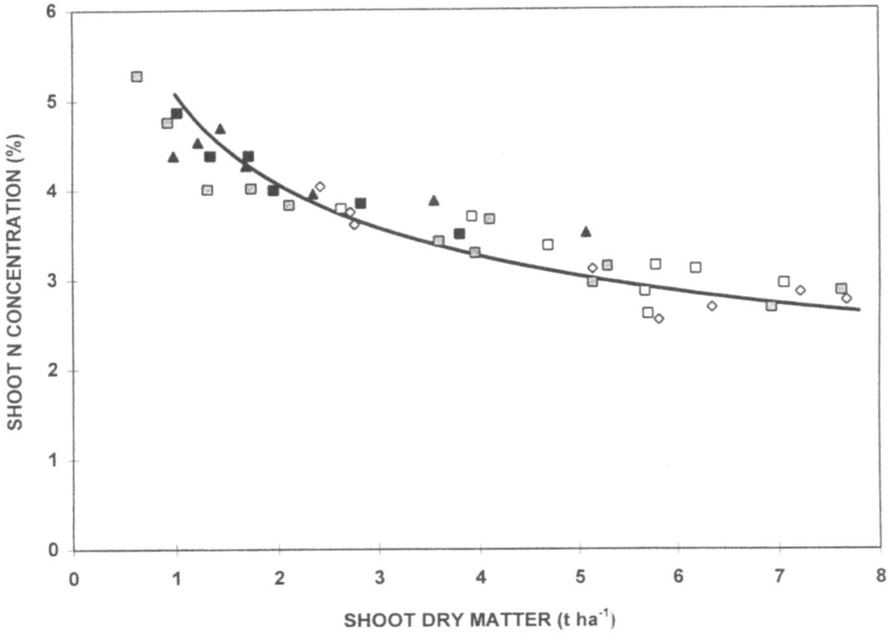

Fig. 6.4. N dilution for other genotypes of pea. Solara: □ 1987; ▩ 1990; ▩ 1992; ■ 1994; other: ◇ Amino 1987; ▲ Alex 1994

earlier described (Fig. 6.3). Comparison between the critical and the observed shoot N% for other genotypes led to the same conclusion (Fig. 6.4). Thus, it can be concluded that N_2-fixating pea crops are able to maintain their plant N% at or near the critical value during all the period of vegetative growth.

6.4
NNI and Yield Determination

Seed number is the most variable yield component in legumes. We investigated for pea the relationship between seed number and NNI at the beginning of seed filling. Previous work showed that seed number is set a few days after this stage (Ney and Turc 1993). NNI was calculated at this stage on experimental plots for Frisson and nonfixating mutants, and seed numbers per unit area were recorded at harvest. One experiment, similar to the experiments in 1990, 1991, and 1993, with various fertilization rates (0, 50, 100, 250, 500 g N m^{-2}), was added in 1995. For NNI < 1, the ratio seed number : seed number of treatments with NNI > 1 was linearly related to NNI at the beginning of seed-filling. For NNI > 1, the maximum seed number was reached (Fig. 6.5).

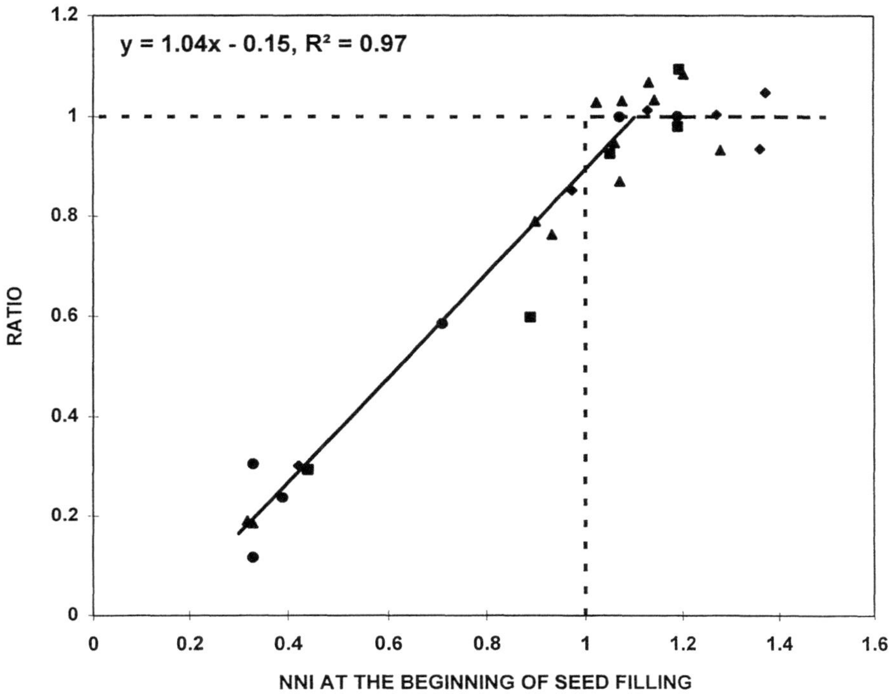

Fig. 6.5. Relationship between NNI at beginning of seed filling and the ratio observed seed number : mean seed number at NNI > 1. ◆ 1990; ▲ 1991; ■ 1993; ● 1995

6.5 Analysis of Yield Variation of Pea Crops in Farm Fields

Previous results have shown that N fertilization regimes had a great influence on the NNI of P2 nonfixating mutants. Yield components are related to the NNI at or near the end of the period in which they are determined. The NNI is therefore a convenient indicator of what occurred with respect of N nutrition; but for crop management, the indicator should detect problems early that can then be corrected. We have seen that seed number was related to NNI at the beginning of seed-filling, but was it related to NNI at the beginning of flowering? Experimental data described previously showed that NNI at the beginning of seed-filling and NNI at the beginning of flowering were very close in value, except, of course for plants that received N fertilizer just after flowering (Fig. 6.6). The slope and intercept of the relationship were not significantly different from 1 and 0, respectively. The relationship between NNI at the onset of flowering and seed number was then established in farm field plots (32 spring-sown fields in 1988, 27 in 1989, and 12 in 1990 in the Seine-et-Marne district, France; Doré 1992). Seed number was recorded at

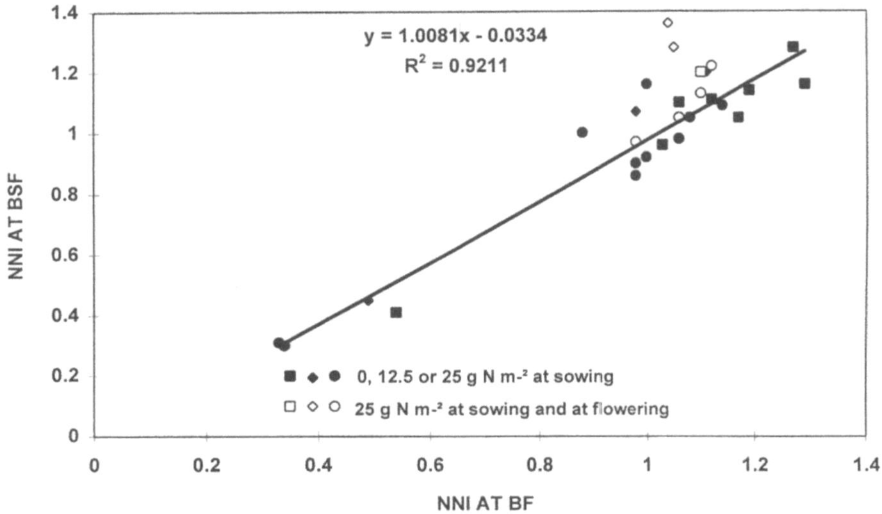

Fig. 6.6. NNI at beginning of flowering (BF) and beginning of seed filling (BSF) in experimental plots. ◆ 1990; ● 1991; ■ 1993

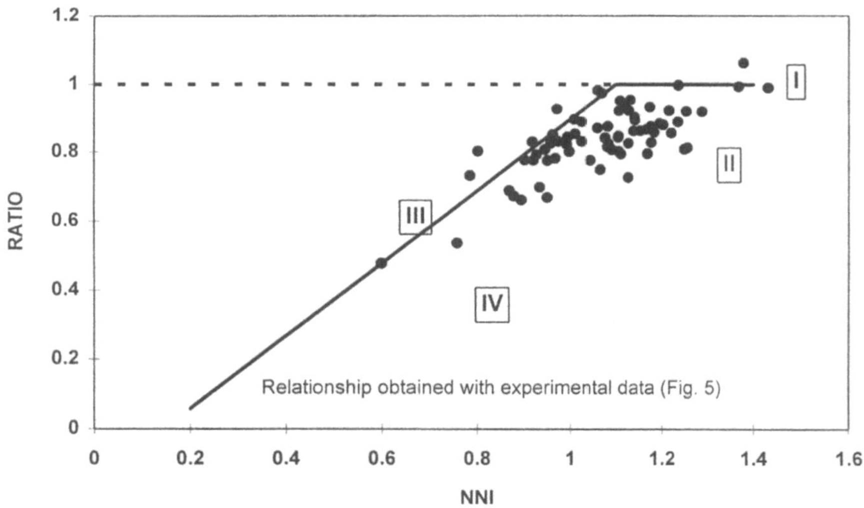

Fig. 6.7. Ratio seed number : maximum seed number and NNI at beginning of flowering in farm fields of pea

harvest and the ratio seed number observed: maximum seed number that year was then calculated. The data obtained in fields were compared to experimental results (Fig. 6.7). Results showed that the experimental relationship obtained with mutants and Frisson corresponds to the upper limit of field points. Some of the points fitted the relation and some of them did not. This result suggests that maximum seed number cannot be obtained with NNI < 1 at the beginning of flowering. On the other hand, many cases with NNI > 1 did not reach the maximum seed number. In these instances, other factors could result in limited yield. From these results, a method could be proposed to separate the effects of nitrogen and the effects of other limiting factors on seed number. In Fig. 6.7, four sets of data are characterized according to NNI and ratio seed number: maximum seed number that year:

1. The first region (I) is characterized by NNI > 1 and maximum seed number.
2. In the second region (II), with NNI > 1 and seed number less than maximum, seed number has been limited by factors other than nitrogen.
3. The third region (III) corresponds to the linear experimental relationship. Nitrogen is the sole limiting factor.
4. In the fourth region (IV) nitrogen and other factors are limiting.

Unfortunately, levels of factors other than nitrogen were not controlled in the farm fields. This method remains to be tested in field conditions under contrasting environmental conditions.

The second question concerning the utilization of NNI for crop management relates to the sensitivity of the method in detecting unfavorable early growth

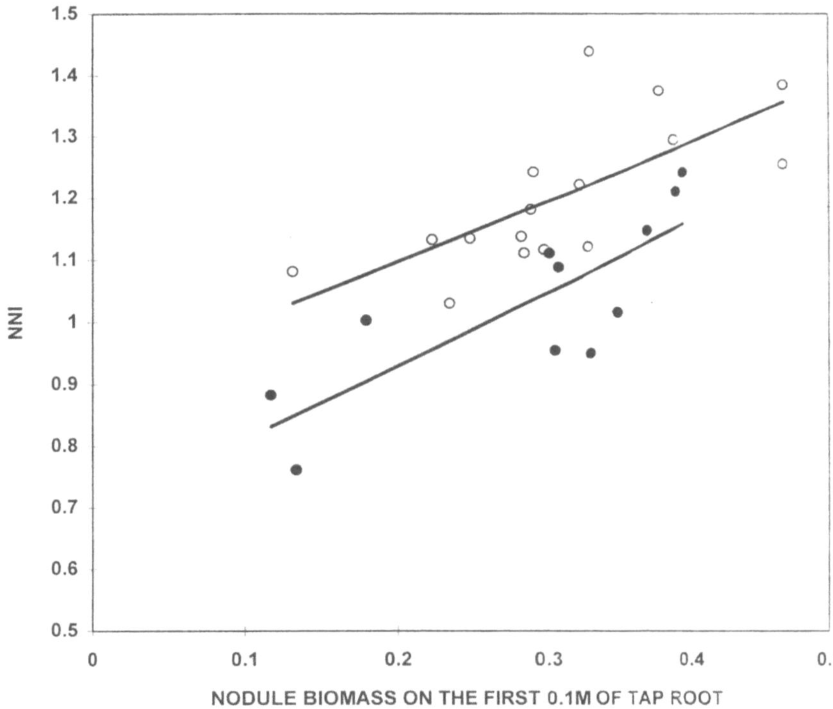

Fig. 6.8. Nodule biomass and NNI at the beginning of flowering in favorable or unfavorable seed bed. ○ Favorable; ● unfavorable

conditions which are known to affect N nutrition. In farm fields, as previously described, soil structure was characterized, nodules were collected on the first 10 cm of tap roots of 20 to 40 plants, and the dry weight of collected nodules was determined. Fields were divided into two groups, according to favorable or unfavorable seed bed structure (thick crust or high proportion of clods with size upper than 10 cm). Results showed that NNI was linearly related to nodule biomass in each group (Fig. 6.8). The greater the nodule biomass, the greater the NNI. The comparison between both groups showed that the slope of the linear relationships was similar, but the intercepts were different. Nodules did not have the same nitrogen fixation efficiency according to soil structure for the same biomass. Thus, NNI is highly sensitive to early growth conditions that affect nodule formation and efficiency.

6.6 Conclusion

Grain legumes behave like other species with respect to dilution of N within the canopy, but only until pods and seeds begin to emerge. After this stage, dilution is compensated for by concentration of N in reproductive organs. This compensation could explain the variability in behavior between legumes. Cultivated as forage, and thus harvested before production of seeds, lucerne behaves like cereals. On the other hand, for early soybean cultivated in the north of France, pods and seeds appear early in the cycle, and the nitrogen dilution curve is not adequate to detect N deficiency. Pea behaves intermediately between these two species. The critical N dilution curve, established with non-nodulating mutants supplied with different N fertilization levels, corresponded approximately to the minimum values obtained for nodulating genotypes in experimental conditions. The coefficients of the equation correspond to those found by Lemaire and Allirand (1993) for lucerne. The critical curve allows the nitrogen nutrition index to be calculated. Experimental results showed that NNI could be related to yield components such as the number of seeds.

Results obtained in farm field conditions are encouraging for the use of NNI as a diagnosis method and crop management tool. The NNI appeared to be sensitive to the early growth conditions of nodule implantation. The relationship between NNI at the beginning of flowering and seed number, which has been established from experimental results, corresponds to the boundary curve for field data. We proposed a method to separate limiting factors other than nitrogen from nitrogen effects. In the absence of control and measurements of these other factors, this method has not been tested.

References

Doré T (1992) Analyse par voie d'enquêtes de la variabilité des rendements et des effets précédant du pois protéagineux de printemps (*Pisum sativum* L.). Thèse de Doctorat de l'INA-PG, Paris, 214 pp

Duthion C, Pigeaire A (1991) Stability of the nitrogen concentration of inoculated white lupins during pod development. Ann Bot 72:55–61

Justes E, Mary B, Meynard JM, Machet JM, Thelier-Huche L (1994) Determination of a critical dilution curve for winter wheat crops. Ann Bot 74:397–407

Lemaire G, Allirand JM (1993) Relation entre croissance et qualité de la luzerne : interaction génotype-mode d'exploitation. Furrages 134:183–198

Munier-Jolain NG (1994) Etude de la variabilité du poids individuel des graines du soja de type indéterminé (*Glycine max* L. Merrill, cv. Mapple Arrow). Thèse de Doctorat de l'INA-PG, Paris, 93 pp

Ney B, Turc O (1993) Heat-unit-based description of the reproductive development of pea. Crop Sci 33:510–514

Sagan M, Ney B, Duc G (1993) Plant symbiotic mutants as a tool to analyse nitrogen nutrition and yield relationship in field-grown peas (*Pisum sativum* L.) Plant Soil 153:33–45

CHAPTER 7

Potatoes

T. Duchenne[1], J.M. Machet[2], and M. Martin[3]

Contents

7.1 Introduction

Nitrogen management is one of the key problems in potato growing, as it affects several parameters of potato growth, including the distribution of dry matter within the plant and the N and dry matter contents of the tubers. Excess of N is as harmful as N deficiency. Therefore, it is necessary to set up a system to determine the N requirement for this plant to avoid the adverse effect of excess N applications on both the crop yield itself, and also the environment. This chapter examines the dynamics of N accumulation and distribution within the plant. An attempt has been made to use the critical plant N% vs. plant mass relationship for this crop despite the fact that tuber filling dominates the plant growth process. The data used in this chapter have been obtained in several experiments, which are summarized in Table 7.1.

[1] Agro-Transfert, Domaine de Brunehaut, 80200 Estrées-Mons, France
[2] INRA Station d'Agronomie, rue Fernand Christ, 02007 Laon Cédex, France
[3] ITCF-ITPT, Domaine de Brunehaut, 80200 Estrées-Mons, France

Diagnosis of the Nitrogen Status in Crops
G. Lemaire (ed)

Table 7.1. Experimental design for the data set used

Year	1993		1994		1995	
Location in France	Eppes (02)		Villers (02)		Villers (02)	
Cultivar	Bintje	Kaptah	Bintje	Kaptah	Bintje	Kaptah
Date of emergence	16-05	17-05	26-05	31-05	22-05	22-05
Plants/m^2	3.6	3.6	3.9	3.9	3.8	3.8
N applications	(1)	(1)	(2)	(2)	(1)	(2)
Samples	5	6	8	9	7	7
Blocks	2	2	3	3	3	3
Plants/blocks	8	8	10	10	30	30

All plots with irrigation.
(1): 0-50-100-150-200-3 × 100 kg N ha^{-1} (ammonium nitrate).
(2): 0-50-100-150-200-3 × 100-3 × 150 kg N ha^{-1} (ammonium nitrate).

7.2 Dry Matter and Nitrogen Distribution During Growth

The distribution of total dry matter (W_t) between the haulm (W_h) and tubers (W_{tu}) of a potato crop shows three phases (Van Heemst 1986). From emergence until the beginning of tuber formation, dry matter accumulates only in above-ground parts (Fig. 7.1). After tuber formation has begun, an increasing fraction of the dry matter is allocated to the tubers, and when the haulm reaches its maximum biomass, all the newly produced dry matter accumulates in the tubers. The maximum accumulation of N in whole plant (N_t) is reached just after the plant has attained its maximum above-ground biomass, although the roots are still able to absorb N later in the cycle (Robinson and Millard 1987; Roberts et al. 1989). By contrast, the amount of N accumulated in the haulm (N_h) reaches its maximum just **before** the maximum aboveground biomass is achieved, and then rapidly decreases, while the amount of N accumulated in the tubers (N_{tu}) continues to increase, indicating that N is transferred from the aerial parts of the plant to the tubers. Millard and MacKerron (1986) indicated that these transfers mainly involve N in the leaves. Estimates of the quantity of N recycled from haulm to tubers have been made for different levels of N supply, indicating an increase in apparent remobilization from 20 to 96 kg N ha^{-1} between low and high N inputs (data not shown).

7.3 Plant N% Decrease with Increasing Plant Mass

Figure 7.2 shows the pattern of decrease in total plant N% with accumulation of dry matter by a potato crop. The use of a log-log scale allows the identification of three phases in the decrease in plant N% with plant mass. The first phase, which is

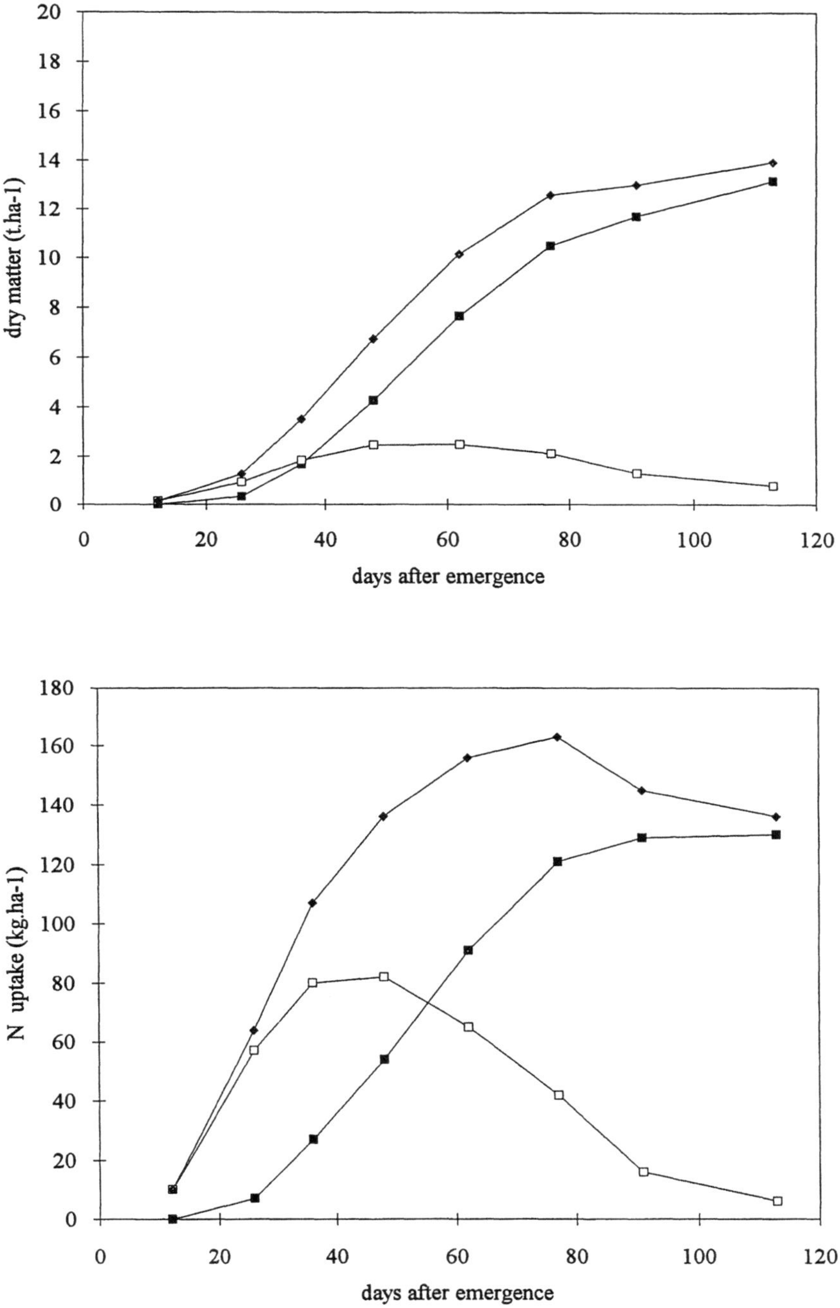

a

b

Fig. 7.1a,b. Dynamics of accumulation of dry matter (**a**) and nitrogen (**b**) by plants after N application of 100 kg ha^{-1} (cultivar Bintje, 1994): in whole plant (◆), in haulms (□), and in tubers (■)

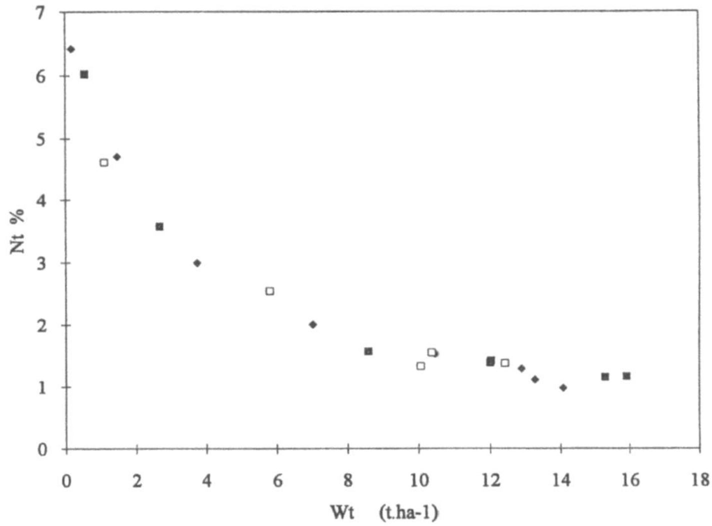

a

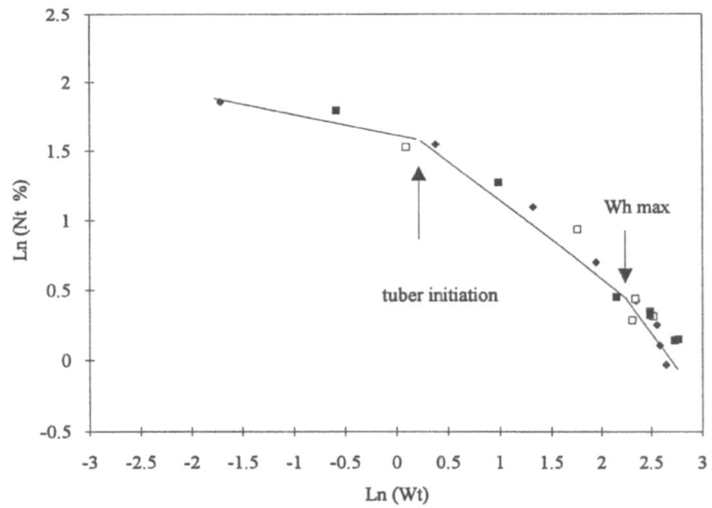

b

Fig. 7.2a,b. Decrease in plant N% with increasing total plant mass for cultivar Bintje after an application of 100 kg N ha^{-1} (1993 ◆; 1994 □; 1995 ■), as normal (**a**) or log-log scale (**b**)

characterized by a slow N% decrease, corresponds to the decline in N% associated with increase in plant size in the situation of isolated plants, as described in Chapter 1. This phase terminates at Ln(Wt) = 0.3, i.e., for a biomass of 1.35 t ha^{-1}, which coincides with the beginning of competition between plants for light and also with the time of tuber initiation. From then on, the rate of decrease of plant

N% with plant mass accelerates as a result of an increasing competition for light between plants (see Chap. 1), and an increasing proportion of the plant mass in the tubers. The third phase starts when the maximum aboveground biomass is reached, and the plant N% decrease is then determined mainly by starch accumulation in the tubers.

7.3.1 Decrease in N% in Haulms and Tubers

The importance of the tubers in dry matter accumulation in potato crops, and therefore in plant N% decrease, would necessitate the addition of a reserve plant compartment as well as the metabolic and the structural compartments. To avoid this complication of the model, it might be possible to analyze the N% decline separately in the haulms and the tubers.

The aboveground biomass of a potato crop can be considered as a canopy to which the theory developed in Chapter 1 can be applied, at least until the maximum accumulated dry matter is reached, i.e., before the transfer of N from leaves to tubers occurs. Figure 7.3a shows that during this period, the decline in N% follows the general pattern of C_3 plants. The N concentration in tubers (N_{tu}%) declines with increasing tuber mass, from 2.0–2.5% at the start of tuber formation to a value of around 1% at harvest (Fig. 7.3b). Several groups (Dyson and Watson 1971; Snyder et al. 1977) have shown that the decrease in the tuber N% corresponds to the transition from the cell multiplication phase to the cell growth and starch accumulation phase. The range of N% values obtained for the different levels of N application (0.8% for 0 kg N ha^{-1} to 1.5% for three dressings of 100 kg N ha^{-1}) is similar to that obtained by Millard (1986).

7.3.2 Determination of Critical N%

Figure 7.4 shows that the maximum possible aboveground biomass was never achieved at any N supply level in the experiment reported here. Thus, the determination of critical N% appears difficult. The reason is that with increasing N supply over the optimum level for total biomass accumulation, tuber formation is inhibited or delayed, and the aboveground biomass continues to accumulate. In these conditions of luxury consumption of N, the maximum total plant mass is achieved, but the haulm continues to grow at the expense of the tubers, leading to a decrease in the harvest index of the crop. Due to the strong effect of N nutrition on the partitioning of dry matter between shoots and tubers, the use of shoot N% for the determination of a critical N concentration in potatoes appears very difficult, and the method must be restricted to only the very early growth period, before tuber initiation. For this reason, we have tried to establish a critical plant N% based on total plant mass.

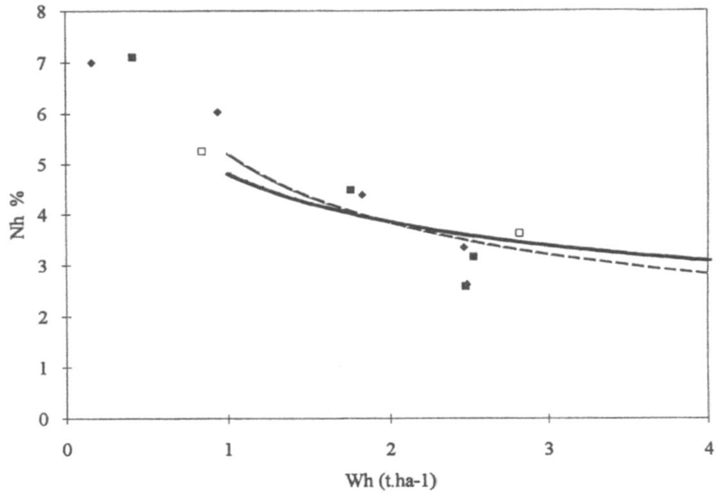

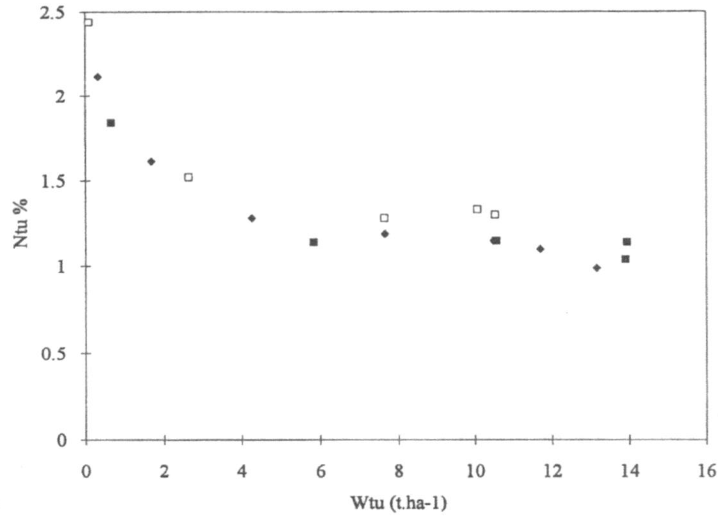

Fig. 7.3. **a** Decrease in N concentration in haulm (N_h%) vs. haulm biomass (W_h) until the maximum aboveground biomass is reached (same symbols as Fig. 7.2). The *two lines* represent the critical curves for wheat: N% = 5.2 $(W_h)^{-0.44}$ (---) (Justes et al. 1994); and for tall fescue: N% = 4.8 $(Wh)^{-0.32}$ (—) (Lemaire and Salette 1984). **b** Decrease in N_{tu}% vs. tuber biomass (W_{tu}) during tuber growth period for the same data set (same symbols as Fig. 7.2)

The most statistically accurate procedure for determining the critical N% is that used by Justes et al. (1994) for wheat. In our case, there were not enough data to use this procedure, so we adopted a simplified one. The critical content has been defined as the lowest plant N% obtained among the group of N treatments having

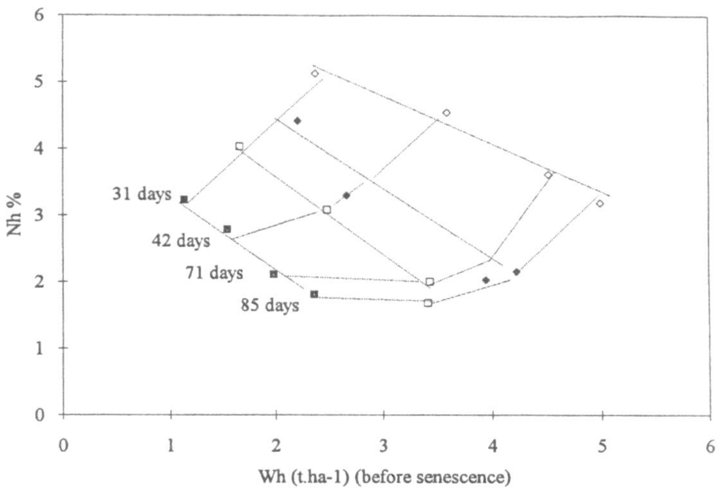

Fig. 7.4. Relationship between N concentration in haulm (N_h%) and haulm biomass for different N applications at different times after emergence, as indicated on the figure. ■ 0; □ 50; ◆ 100; ◇ 3 × 100 kg N ha^{-1} (cv. Kaptah Vandel)

the greater dry matter production, after classification of the dry matter values by analysis of variance, comparing the means in pairs using the Newman and Keuls test for $P = 0.10$. The N% and the dry matter value of these treatments define the coordinates of a critical point. A total of 17 data points were obtained for the cultivar Bintje and 19 for Kaptah Vandel (Fig. 7.5).

In line with the theory developed in Chapter 1, several models relating critical plant N% (N_c) to plant mass (W_t) were fitted to these points:

$$N_c = aW_t^{-b} \tag{1}$$

$$N_c = N_s + aW_t^{-b} \tag{2}$$

$$N_c = N_s + (N_m - N_s)(W_t/W_i)^{-b} \tag{3}$$

$$N_c = N_s + (N_m - N_s)[x/(x + W_t)] \tag{4}$$

(see Chap. 1 for the significance of the coefficients).

Equations (2) and (3) were found to give the same coefficients of determination as Eq. (1) (Table 7.2). Equation (4) gave a better mean squares error and distribution of residuals. N_s represents in this case the global N% of the structural + reserve compartment. Its adjusted value is zero for Eqs. (3) and (4) [except for Bintje in Eq. (4)]. This estimation of a value of N_s not significantly different from zero indicates that the size of the structural + reserve plant component is mainly driven by starch accumulation, i.e., with negligible N inputs.

For simplification, Eq. (1) can be used for the determination of critical plant N%. The difference between the two varieties used does not appear statistically significant. Thus, a common equation can be adopted for the determination of the critical plant N%;

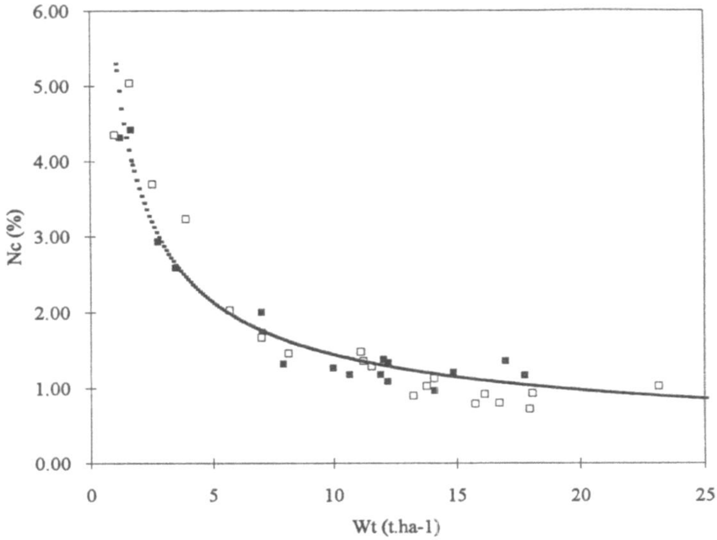

Fig. 7.5. Critical plant N% vs. total plant mass for the two cultivars, Bintje (■) and Kaptah Vandel (□). The equation of the fitted curve is N% = 5.21 $(W_t)^{-0.56}$

$$N_c = 5.21(W_t)^{-0.56}.$$

Using this equation, a nitrogen nutrition index (NNI) can be calculated as the ratio of observed (N_{obs}) to critical (N_c) plant N%, as proposed in Chapter 1.

Figure 7.6 gives the relationship between tuber and total biomass for the two varieties when plant N% is maintained at the critical values. The curves are linear for values of total biomass greater than 5 t ha^{-1}:

$$\text{for Bintje: } W_{tu} = 1.09(W_t - 2.95) \; r^2 = 0.98$$

$$\text{for Kaptah Vandel: } W_{tu} = 1.04(W_t - 3.76) \; r^2 = 0.96.$$

The difference in the value of the intercept on the x axis between the two varieties indicates that tuber growth is delayed in Kaptah Vandel in comparison to Bintje; an extra plant mass of 0.8 t ha^{-1} is needed for the former to achieve similar tuber mass.

7.4 Effects of Nitrogen Nutrition Status on Yield

Yield of a potato crop (W_{tu}) can be described as the product of the quantity of intercepted PAR, the radiation use efficiency (RUE), and the harvest index (HI):

$$W_{tu} = PAR*RUE*HI$$

Table 7.2. Comparison of models for decline of critical plant N% with increasing plant mass

Models	Parameters	Bintje			Kaptah Vandel			Bintje and Kaptah		
			MSE	r2		MSE	r2		MSE	r2
1	a	5.17	0.054	95.6	5.25	0.201	93.9	5.21	0.128	96
	b	0.56			0.56			0.56		
2	N_s	0	0.058	95.6	0	0.214	93.9	0	0.133	96
	a	5.17			5.25			5.21		
	b	0.56			0.56			0.56		
3	N_s	0	0.062	95.6	0	0.228	93.9	0	0.137	96
	N_m	4.57			4.64			4.61		
	W_o	1.24			1.25			1.25		
	b	0.56			0.56			0.56		
4	N_s	0.5	0.050	96.3	0	0.103	96.6	0.03	0.085	97
	N_m	8.41			6.77			6.63		
	x	1.27			2.74			2.66		

MSE: mean squares error.
1 $N_c = aW_t^{-b}$.
2 $N_c = N_s + aW_t^{-b}$ and $N_s \geq 0$.
3 $N_c = N_s + (N_m - N_s)(W_t/W_o)^{-b}$ and $N_s \geq 0$.
4 $N_c = N_s + (N_m - N_s)[x/(x - W_t)]$ and $N_s \geq 0$.

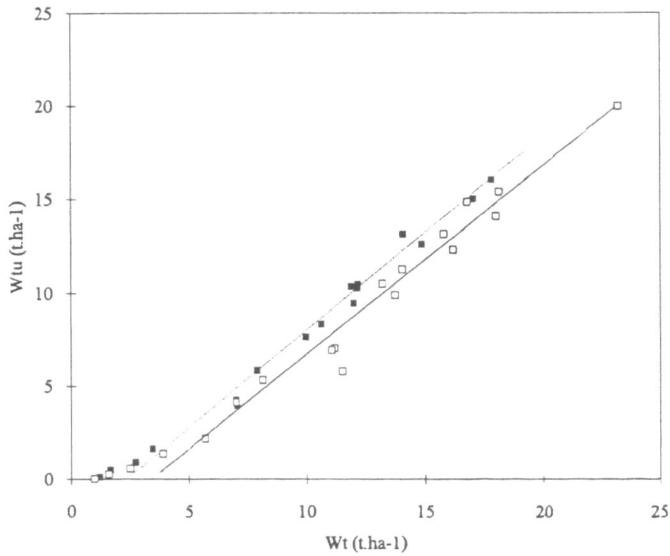

Fig. 7.6. Relationship between tuber and total biomass for the same data set as in Fig. 7.5 (same symbols)

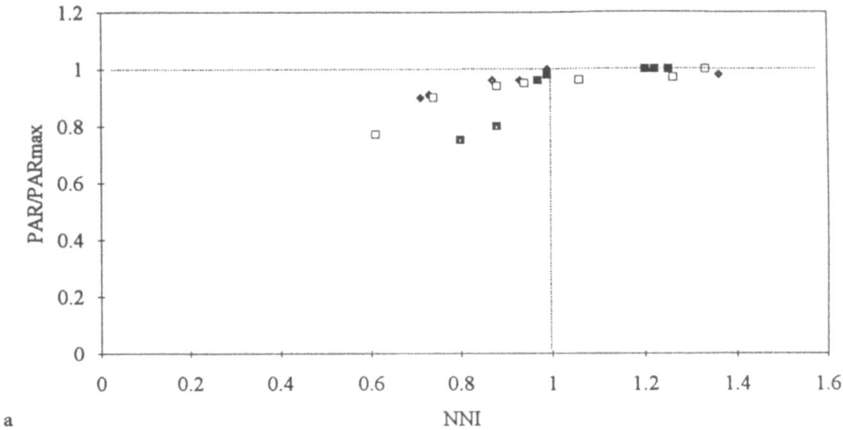

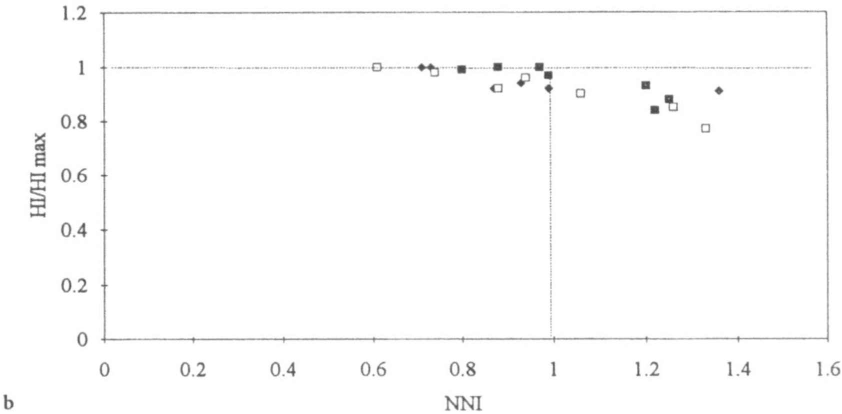

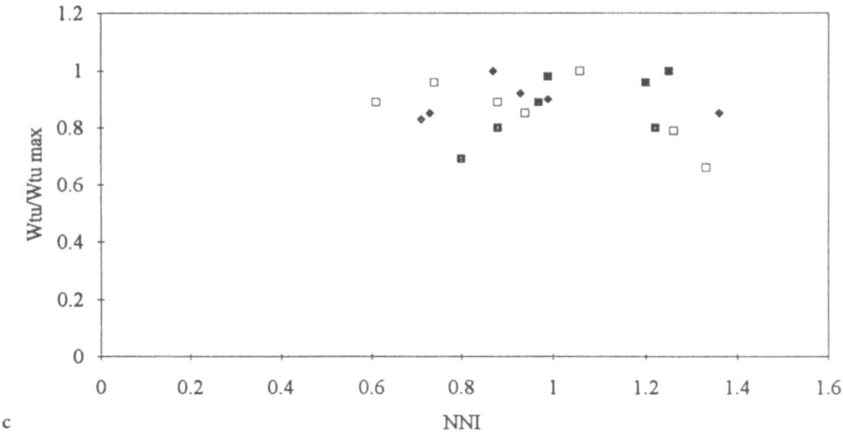

Fig. 7.7a–c. Relationships between the nitrogen nutrition index (NNI) and **a** the relative amount of intercepted PAR, **b** the relative harvest index (*HI*), and **c** the relative yield (*Wtu*) for the cultivar Kaptah Vandel in 1993 (◆), 1994 (□), 1995 (■)

The effects of N nutrition on yield can be analyzed through the relationships between the NNI and the reduced variables, PAR/PAR_{max}, RUE/RUE_{max}, and HI/HI_{max}, calculated at harvest. NNI has opposite effects on the PAR/PAR_{max} and HI/HI_{max} ratios; a decrease in NNI below the optimum value of 1 decreases the amount of intercepted PAR (Fig. 7.7a), while the harvest index is not affected (Fig. 7.7b). On the other hand, a value of NNI greater than 1 is associated with a decrease in the harvest index. Thus, both deficiency and excess of N lead to a reduction in yield (Fig. 7.7c). We found that the RUE was not altered by N deficiency in potatoes, in agreement with the findings of Millard and Marshall (1986). Such opposing effects of N nutrition on yield have been reported by several authors (Millard and Marshall 1986; Biemond and Vos 1992; Vos and Biemond 1992; Vos 1995). Tuber initiation by plants grown in nutrient solutions with photoperiods and temperatures that favor tuber formation can be totally inhibited by the input of N (Krauss 1978, 1985; Sattelmacher and Marschner 1979). In the field, the date of tuber initiation does not seem to be affected. This is unlike the date at which rapid tuber growth begins, which is retarded by N application, allowing the shoot to continue to accumulate dry matter.

7.5 Conclusion

The decrease in plant N% in the whole plant of potato during growth is mainly driven by the accumulation of N-poor dry matter in the storage organs, that is, the tubers. There are three growth phases: before $1.35\,t\,ha^{-1}$, the plant mass is essentially composed of leaves and stems, and the decrease in critical plant N% follows the common pattern of C_3 crops; between $1.35\,t\,ha^{-1}$ and the maximum aboveground biomass, tuber formation occurs in tandem with leaf and stem growth; and finally when the maximum aboveground biomass is reached, all dry matter produced is allocated to the tubers, which leads to an acceleration of the decline in whole-plant N%.

Because of changes in the distribution of dry matter caused by N supply, it appears difficult to establish separate critical curves for the aerial plant parts and the tubers. Hence, the critical curve for plant N% vs. plant mass at the whole plant level must be used for the diagnosis of the N nutrition status of the crop. There was no difference between the two varieties Bintje and Kaptah Vandel, and an N nutrition index could be calculated for both cultivars. Deficiencies of N ($NNI < 1$), cause a decrease in the amount of radiation intercepted, while excess of N ($NNI > 1$) leads to a decrease in the harvest index. Hence, maximum tuber yield is achieved only under conditions of N nutrition that are maintained narrowly around the optimum. Thus, an early diagnosis of the N nutrition status of potato crops appears to be of great interest, in order to manage very carefully the N fertilization of the crop, to prevent both N deficiency and excess of N, and so to optimize yield.

References

Biemond H, Vos J (1992) Effects of nitrogen on the development and growth of the potato plant. 2. The partitioning of dry matter, nitrogen and nitrate. Ann Bot 70:37–45

Dyson PW, Watson DJ (1971) An analysis of the effects of nutrient supply on the growth of potato crops. Ann Appl Biol 67:47–63

Greenwood DJ, Neeteson JJ, Draycott A (1985) Response of potatoes to N fertilizer: quantitative relations for components of growth; dynamic model. Plant Soil 85:163–183

Greenwood DJ, Lemaire G, Gosse G, Cruz P, Draycott A, Neeteson JJ (1990) Decline in percentage N of C_3 and C_4 crops with increasing plant mass. Ann Bot 66:425–436

Justes E, Mary B, Meynard JM, Machet JM, Thelier-Huche L (1994) Determination of a critical nitrogen dilution curve for winter wheat crops. Ann Bot 74:397–407

Krauss A (1978) Tuberization and abcissic acid content in *Solanum tuberosum* as affected by nitrogen nutrition. Potato Res 21:183–193

Krauss A (1985) Interaction of nitrogen nutrition, phytohormones and tuberization. In: Li PII (ed) Potato physiology. Academic Press, New York, pp 209–230

Millard P (1986) The nitrogen content of potato (*Solanum tuberosum* L.) tubers in relation to fertilizer nitrogen application. The effect on amino acid composition and yields. J Sci Food Agric 37:107–114

Millard P, MacKerron DKL (1986) The effects of nitrogen application on growth and nitrogen distribution within the potato canopy. Ann Appl Biol 109:427–437

Millard P, Marshall B (1986) Growth, nitrogen uptake and partitioning within the potato (*Solanum tuberosum* L.) crop, in relation to nitrogen application. J Agric Sci Camb 107:421–429

Roberts S, Cheng HH, Farrow FO (1989) Nitrate concentration in potato petioles from periodic applications of 15N-labelled ammonium nitrate fertilizer. Agron J 81:271–274

Robinson D, Millard P (1987) Short-term uptake rates and partitioning of nitrogen in a potato crop. J Exp Bot 38:841–848

Sattelmacher B, Marschner H (1979) Tuberization in potato plants as affected by application of nitrogen to the roots and leaves. Potato Res 22:49–57

Snyder J, Desborough S, Holm D (1977) Accumulation of protein, non-protein nitrogen and starch during tuber growth of three potato cultivars. Am Potato J 54:545–555

Van Heemst HDJ (1986) The distribution of dry matter during growth of a potato crop. Potato Res 29:55–66

Vos J (1995) Nitrogen and the growth of potato crops. In: Haverkort AJ, MacKerron DKL (eds) Potato ecology and modelling of crops under conditions limiting growth. Current issues in production ecology. Kluwer, Dordrecht, pp 115–128

Vos J, Biemond H (1992) Effects of nitrogen on the development and growth of the potato plant. 1. Leaf appearance, expansion growth, life spans of leaves and stem branching. Ann Bot 70:27–35

CHAPTER 8

Mixed Crops

P. Cruz[1] and J.F. Soussana[2]

Contents

8.1 Introduction

A mixed crop may be defined as any plant stand where two or more species grow together in the same field, during the whole or part of their growth cycles (Baldy and Stigter 1993). This broad definition covers a large range of cropping conditions and crop configurations, from the simplest clover-grass sward to the complex agroforestry systems. In addition to any specific advantage which might justify the use of a particular mixture, there is one main reason for cultivating mixed crops; they are believed to use environmental resources more efficiently than monocultures (Stigter and Baldy 1995). In the case of light and water, such greater resource use efficiency could be explained by the fact that the above- and belowground exchange organs (leaves and roots) of the associated species explore different volumes of air and soil, or differents ecological niches (Vandermeer 1989). Under these circumstances, total radiation interception and transpiration by a mixed crop could be higher than those of corresponding monocultures (Tournebize and Sinoquet 1995; Tournebize et al. 1996). This is in agreement with the lower relative value of interspecific competition compared to intraspecific competition which is often reported (see Haynes 1980). However, with N, which is mainly localized in

[1]INRA Unité Agropédoclimatique de la Zone Caraïbe, BP 515, 97165 Pointe-à-Pitre Cédex, Guadeloupe, France; Present address: INRA Station d'Agronomie, BP 27, 31326 Castanet-Tolosan Cédex, France
[2]INRA Station d'Agronomie, 12 Avenue du Brézet, 63039 Clermont-Ferrand Cédex 02, France

Diagnosis of the Nitrogen Status in Crops
G. Lemaire (ed)

the upper layer of the soil and, therefore, in principle, more subject to direct competition, its improved capture by a mixed crop could be related to another factor. Most mixed crops are grass-legume mixtures, which implies that two different sources of N can be used by the crop: uptake from the soil by grass and legume and legume symbiotic N_2 fixation from the atmosphere. Isotopic methods, using either a fertilizer labeled with ^{15}N (Vallis et al. 1967) or direct labeling by $^{15}N_2$ (Russelle et al. 1994), and enzymatic techniques (Soussana et al. 1989) may be used to investigate this.

It has been reported that in grass-legume mixtures the benefit for the associated grass could have a double origin. Firstly, due to symbiotic N_2 fixation by the legume, its competitiveness for soil nitrogen is usually lower (Soussana et al. 1989; Faurie 1994), which leads to an increased availability of soil N for the grass (Vallis et al. 1967). Secondly, a belowground transfer of N could occur from the legume to the grass, by either direct (Virtanen et al. 1937; Ta et al. 1986) or indirect pathways (Dubach and Ruselle 1994). The corresponding advantage for the N nutrition of the mixed grass would be greater in situations where the N availability of the soil is low.

The evidence for a better N nutrition of the mixed grass often came from measurements of a higher N% in the mixed grass than in the control grass monoculture (Sangakkara and Attanayake, 1993). However, the possible advantage of mixing legumes and grasses with respect of N nutrition cannot be assessed when comparison is only based on the N concentration of the shoot. We know that N% is strongly correlated to the dynamics of dry matter accumulation in the crop (see Lemaire and Gastal, Chap. 1, this Vol.). Therefore, a higher N% could result from a reduced growth of the mixed grass, rather than from a greater N availability. Thus, to conclude about any hypothetical advantage of mixed grasses, it is necessary to demonstrate that their N nutrition is improved by the associated legume. As discussed in Chapter 1, by plotting N concentration against accumulated dry matter, it is possible to estimate the level of N nutrition of a crop. This method,

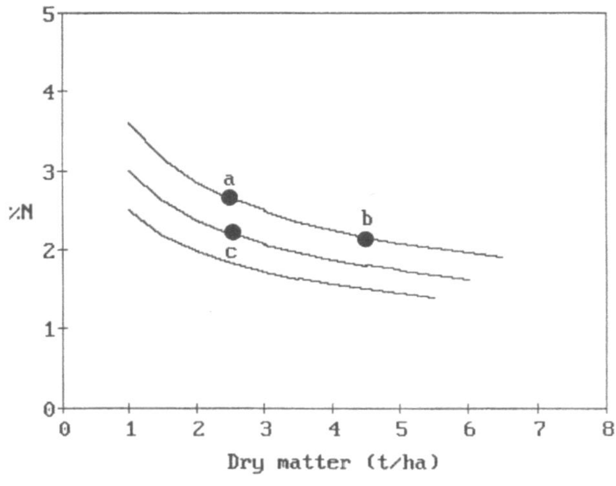

Fig. 8.1. Changes in N concentration dues to: *b* → *a* light competition; *a* → *c* N competition; *b* → *c* light and N competition

usually applied to monocultures, could be used in the same way to analyze changes in the N status of a grass species when cultivated in a mixed crop. Moreover, the use of this approach of comparing pure and mixed stands allows us to analyze separately the competitive relationship for light and nitrogen resources. Figure 8.1 shows how an increase in the N% in a mixed species could be due to competition for light (reduced growth) or to a higher N availability. However, this method does not allow us to discriminate between the two pathways of legume N nutrition, nitrate uptake, and N_2 fixation.

The application of critical N% curves to mixed crops has to take into account specific aspects of these stands relating to plant density.

8.2 Plant Density Arrangements in Mixed Crops

It was previously discussed (Chap. 1) how the relationship between the shoot N% and the cumulative dry matter in a dense crop is a function of the light microclimate within the canopy. Such a relationship has been well documented for diverse monospecific stands (see Chaps. 1 and 2). The decline in leaf N% due to the reduction of irradiance within the canopy has also been observed for the components of a mixed crop (Soussana et al. 1995). The light microclimate in any crop is highly dependent on the plant density, so the decline in N% during growth is less in isolated plants than in plants in a dense stand (Fig. 1.11 in Chap. 1). In isolated plants, the decline in N% is due only to a decrease in the ratio between metabolic and structural tissues which have different N concentrations. Thus, the critical N% for potential growth of a crop differs according to the density at which the stand grows. In intercropping experiments, changes in the N nutrition level of a component in a mixture need to be compared with the corresponding pure stand. With most agronomic experiments, it is usually assumed that the monocultures used for this comparison have the optimal density, that is, the density allowing the highest yield (Vandermeer 1989). Nevertheless, in some experiments, the pure stand used as a control has a lower plant density than the optimal monoculture (for instance, double-spaced rows in additive designs), and its critical N% will be intermediate between that of an isolated plant and that of a dense stand. This lower density of the control pure crop may create a bias in studies where the aim is to evaluate interspecific competition. This example shows that the application of the critical N% approach to evaluating the N status of mixed crops has to take into account the methodological constraints posed by the density variation of the control treatment.

8.3 Experimental Designs and Some Application Examples

The simplest layout to study mixed crops is the cultivation of components in alternate rows. In this case, variations between simple- and double-spaced rows in

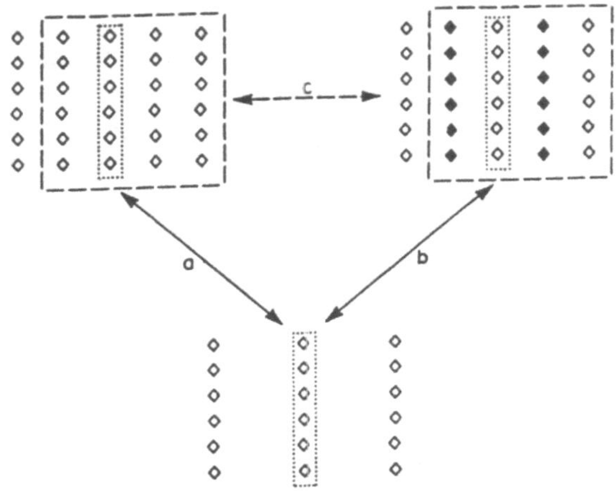

Fig. 8.2. Layout allowing the study of: → *a* intraspecific competition (row basis); → *b* interspecific competition (row basis); → *c* balance between intra and interspecific competition (row or area basis)

pure crops allow the separation of intra- and interspecific competition and the balance between them when compared to the mixed crop (Fig. 8.2). Such a design allows an analysis of dry matter to be measured on a row basis or on an area basis, depending on the aim of the study.

8.3.1 Additive Experiments

In additive experiments, the density of species A is the same in the mixture and in the pure stand. When species B is added, this increases the total density of the stand. The additive design is well adapted to study the effect of interspecific competition, i.e., the effect of the presence of B on the N nutrition of A, and vice versa. In this case, the accumulated dry matter of the species A may be expressed both per row and per unit area. Furthermore, if the density of the species A is that of an optimal pure crop, the critical N% may be compared to the curve describing the optimal relationship between dry matter and N% for this species, or for its metabolic group (C_3 or C_4). This kind of comparison was made by Cruz et al. (1991), studying tall fescue in a pure sward or associated with white clover. Both the mixed and pure tall fescue were sown at the same optimal density, with the white clover added to create the mixed grass-legume sward. In this way, assuming that the tillering of the grass in the mixture was not depressed by clover, interspecific competition was added to the intraspecific competition. Nevertheless, a positive effect of the presence of the legume on the N nutrition of the grass was observed under N-limiting conditions (Fig. 8.3). When ample N fertilizer was supplied, the positive effect of the legume disappeared, and both pure and mixed tall fescue were very similar and almost reached the critical N% described for the potential growth of C_3 species. Nevertheless, with clonal plants like grasses and clovers there is no direct link between the sowing density and the final density of

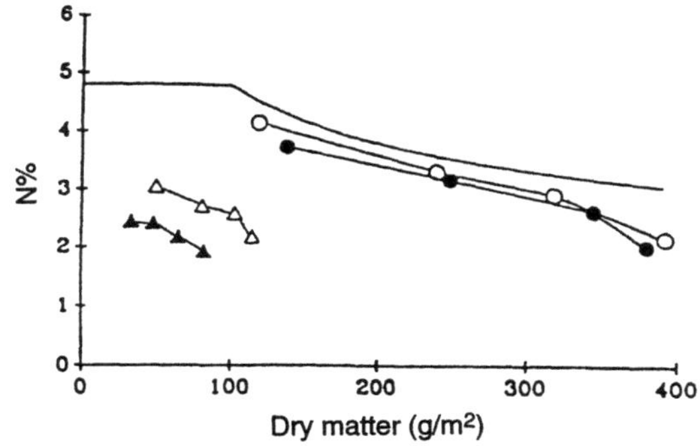

Fig. 8.3. N concentration of *Festuca arundinacea* in pure stand or mixed to *Trifolium repens*, as a function of grass dry matter accumulation. Pure stand with 150 (●) and 0 (▲) kg N ha^{-1}; mixed stand with 150 (○) and 0 (△) kg N ha^{-1}; *solid line* critical N% reference for C_3 species (from Cruz et al. 1991)

the plants. Hence, adding a clover row might have resulted in a decrease in the grass tiller density, thereby affecting the intraspecific competition.

A common case of additive experiments is the sylvopastoral system, where trees or shrubs are planted into a perennial sward, Under tropical conditions, the woody component is usually a legume species, which is intended to enrich the system by fixing N. Critical N% curves used to describe the N status of the underlying grass showed that its level of N nutrition was clearly improved (Cruz et al. 1993). Using original data from Belsky (1992), it is possible to confirm this positive effect of the trees on the N nutrition, because both the biomass and the N% of the grass were enhanced. However, in another study (Puerto Martin et al. 1984), a different conclusion has to be reached, because the positive effect of the trees on the N concentration of the grass was due to reduced growth of the latter. Wilson (1990) suggested that the occasional enhanced availability of N in sylvopastoral stands is due to a reduction in water stress by shade, and not to a direct enhancement of N availability due to the legume N fixation. Whatever the origin of this discrepancy in the literature, the effect of the mixture on the N nutrition can be assessed by plotting N% against biomass.

8.3.2 Substitutive Experiments

In substitutive experiments, the monocultures and the mixtures are sown at the same total density, and usually each of the two mixed species is sown at half this total density. Such a layout results in substitution of part of the intraspecific competition by interspecific competition, and therefore they cannot be analyzed separately. This is what classically occurs in agronomical experiments carried out in the field, where the balance between intra- and interspecific competition is analyzed. Substitutive designs have also frequently been used in analytical experiments where the proportion of both species are varied, while the total density of the crop is maintained as constant. These "replacement series" were studied for grass-legume mixtures by de Wit et al. (1966). However, in most agronomical studies, only a part (normally half) of the plant population of the pure stand is replaced in the mixed stand by the other component, sown at half the density of the monoculture.

Critical N% curves have been applied to a large range of substitutive designs with species of contrasting growth habits. For instance, Marín et al. (1995) established an inter-intraspecific competition balance in a sorghum-Jack bean (*Canavalia ensiformis*) mixture. They attributed the better N nutrition of the mixed sorghum to the greater availability of soil N. The fixation activity of the legume was high, so it was presumably less competitive for soil N than the grass. As a consequence, the intraspecific competition for N in the sorghum was greater than the interspecific competition. By comparing only the N% of the mixed and pure sorghum, it could be concluded in this experiment that the mixture had no effect. In fact, the N% of the mixed and pure sorghum was similar at the beginning and at the end of the growth cycle (first and last points in Fig. 8.4), but differences clearly occurred in between. The dynamics of the changes in N% should therefore be analyzed during the growth cycle, avoiding the use of instantaneous data.

In another substitution experiment, Soussana and Arregui (1995) replaced half a stand of an erect grass species, *Lolium perenne*, by a stoloniferous legume, *Trifolium repens*. Spreading species such as white clover and creeping grasses need special treatment, because they do not maintain the row structure for a long time and may rapidly colonize the interrow. In this case, it is not possible to express the dry matter of each species on a row basis, and comparisons between pure and mixed stands have to be made at the whole crop level, without the possibility of discriminating between inter- and intraspecific competitions. In their work, Soussana and Arregui (1995) pointed out the inconvenience of expressing the critical N% of the mixed grass using biomass data on an area basis, because the area of a row in the mixed crop cannot be assessed. They suggested, and compared,

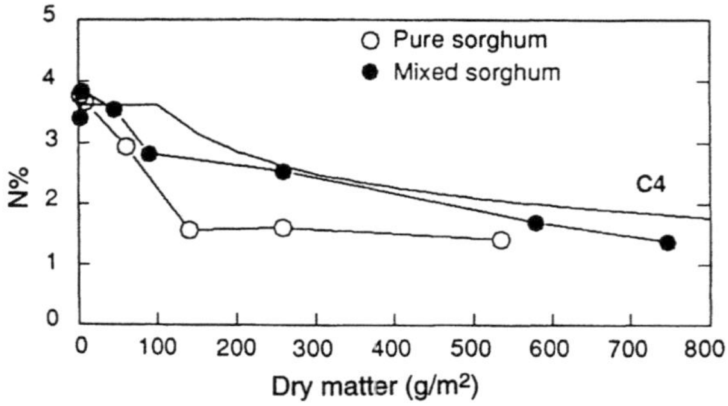

Fig. 8.4. N concentration of *Sorghum vulgare* in pure stand or mixed to *Canavalia ensiformis*, as a function of dry matter accumulation. *Solid line* Critical N% reference for C_4 species (from Marín et al. 1995)

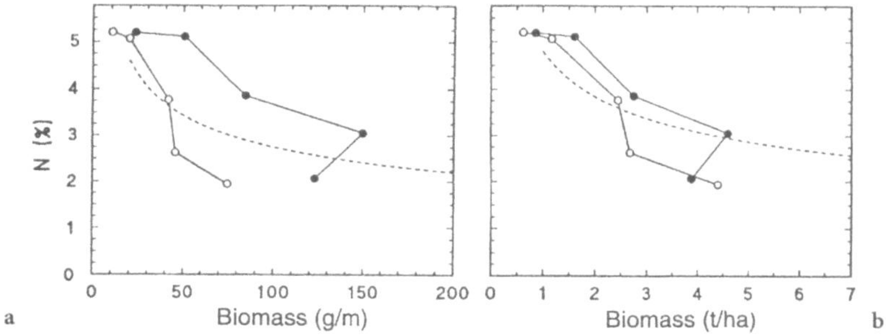

Fig. 8.5a,b. N concentration of *Lolium repens* in pure stand or mixed to *Trifolium repens*, as a function of the grass component (**a**) or the whole stand (**b**) dry matter accumulation. ● Mixed grass; ○ pure grass; *dotted line* critical N% reference for C_3 species (from Soussana and Arregui 1995)

two different ways to analyze the expected changes in the N nutrition of ryegrass when associated with white clover.

The first approach consists of plotting the N% of pure and mixed grasses against their corresponding biomass on a row basis. Presenting data on a row basis prevents them from being compared to the critical N% curve used for optimal growth of C_3 perennials, previously described in Chapter 1 (this Vol.). To make this comparison possible, these authors proposed a correction of the original equation to express it on a row basis (Fig. 8.5a). The resultant curve cannot be generalized, as it varies with the spacing between rows, and should therefore be adapted to each particular case.

The second proposed procedure to compare pure and mixed ryegrass consists of plotting the N% of the mixed grass against the total biomass of the mixture (grass + legume), as shown in Fig. 8.5b. The reason for this is simple. If the decrease in N% in the plant during growth is mainly determined by competition for light within the canopy (Lemaire and Gastal, Chap. 1, this Vol.), the biomass of the neighboring species has to be taken into account because it contributes to the canopy light capture and to light attenuation. Figure 8.5 shows how the use of the grass biomass on a row basis (first procedure) enlarges the difference between both grass stands and overestimates the critical N% of the mixed grass, when compared to the reference curve. Nevertheless, in both cases, the N nutrition of the ryegrass was improved by mixing with the white clover. This was also interpreted as resulting from the lower density of the mixed grass and the reduced competitiveness of the legume for soil N, which benefited the ryegrass in the mixture. However, transfer of fixed N from the clover to the grass may also have contributed to the improved N nutrition of the grass in the mixture.

The procedure using the total biomass of the mixture to generate the N% curve of only one of the mixture components could be applied to reevaluate data from another subtitutive experiment carried out on a mixed pasture (Cruz and Sinoquet 1994). In this trial, Pangola grass (*Digitaria decumbens*) was grown associated, or not, with a tropical perennial legume, *Arachis pintoi*. Both species have a stoloniferous growth habit, and after being sown in rows (alternate rows in mixed stands), spread quickly and colonize the whole plot. In the original paper, the biomass of pure and mixed grass was expressed on an area basis because both were considered as being homogeneously distributed in the plots. It was concluded that the N nutrition of the grass was the same in both stands, whatever the N supply (0 or 150 kg ha^{-1}). Therefore, the mixed legume was not able to improve the N nutrition of the Pangola grass (Fig. 8.6a).

However, if the N% of mixed Pangola is plotted against the total biomass of the mixture, as proposed by Soussana and Arregui (1995), different conclusions are obtained (Fig. 8.6b). When no N was supplied, the Pangola grass had a higher N% in the mixture than in the pure crop for an identical biomass. When a large amount of N was supplied, the legume was almost suppressed and the biomass of the mixed stand (grass + legume) was barely greater than that of the mixed grass. As a consequence of the large supply of N fertilizer, both mixed and pure grass stands had the same level of N nutrition, close to the critical N% described for C_4 species by Lemaire and Chartier (1992).

Yet the difference observed between the grass stands without N supply could be smaller if we consider that, in the mixture, the legume layer is partly overtopped by the grass foliage. In this case, *Arachis* is probably less competitive for light than this kind of approach assumes. To avoid the uncertainty about the partitioning of radiation between the species in the mixture and its effect on the expression of the critical N% curves, it could be possible to plot the N% data against the total radiation intercepted by each component. In the preceding example, this analysis was possible because light partitioning between species, which cannot be measured, was simulated in the original paper. Figure 8.7 shows that, whatever the N supply, an identical relationship exists for mixed and pure Pangola grass. This confirms the lack of a legume effect on the N nutrition of the grass, which was shown previously by using the grass biomass to generate the critical N% curves.

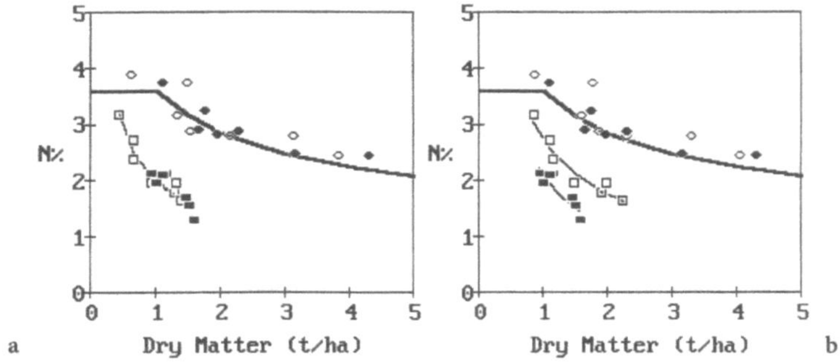

Fig. 8.6a,b. N concentration of *Digitaria decumbens* in pure stand or mixed to *Arachis pintoi*, as a function of the grass component (**a**) or the whole stand (**b**) dry matter accumulation. Pure stand with 150 (●) and 0 (■) kg N ha^{-1}; mixed stand with 150 (○) and 0 (□) kg N ha^{-1}; *solid line* critical N% reference for C_4 species

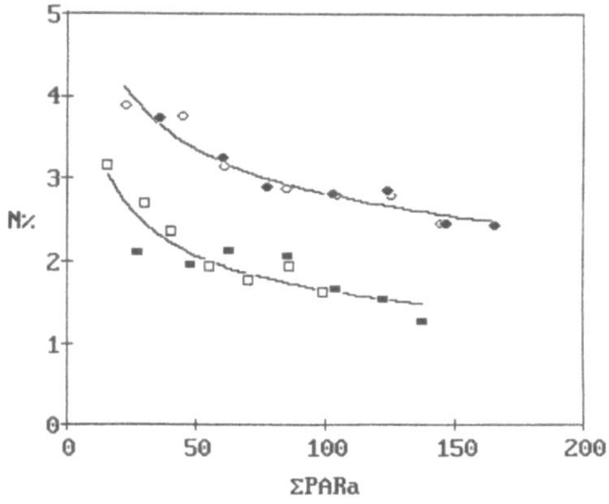

Fig. 8.7. N concentration of *Digitaria decumbens* in pure stand or mixed to *Arachis pintoi*, as a function of the calculated total PAR absorbed by the grass. *Symbols* identical to Fig. 8.6

In any case, these different results show the difficulties encountered when the critical N% concept is applied to a mixed crop which includes spreading species, as this trait modifies the density, and hence the N% of the shoots, in a manner that cannot be controlled by the experimental design.

8.3.3
Integrated Additive-Substitutive Experiments

With erect species, the accumulated dry matter may be expressed on either a row or an area basis, depending upon the aim of the study and the layout of the experiment. With these species, a complete experimental design including both additive and substitutive arrangements is possible. An example of this was the study of inter- and intraspecific relationships between alfalfa (*Medicago sativa*) and cooksfoot (*Dactylis glomerata*) carried out by Cruz and Lemaire (1986). In this trial, both species were sown in single- and double-spaced rows for pure crops and in alternate single-spaced rows for the mixture (see scheme in Fig. 8.2). Such a layout gives the possibility of analyzing the role of the interspecific competition for N and for light (taking the half-density monocultures as a control), as well as the balance between intra- and interspecific competition for agronomic conclusions (taking the high-density monocultures as control).

In this work, no beneficial effect from the legume on the N nutrition of the grass was demonstrated, either with or without N fertilizer supply. The higher N% of mixed cooksfoot was not due to an improvement in the N availability, but to reduction in the grass growth when mixed with alfalfa (Fig. 8.8). This result was previously interpreted as strong competitiveness of alfalfa for soil N. However, this conclusion should be reconsidered, as according to the light-N linkage at the canopy level, the poorer N nutrition of the mixed grass could also result from competition for light by the alfalfa.

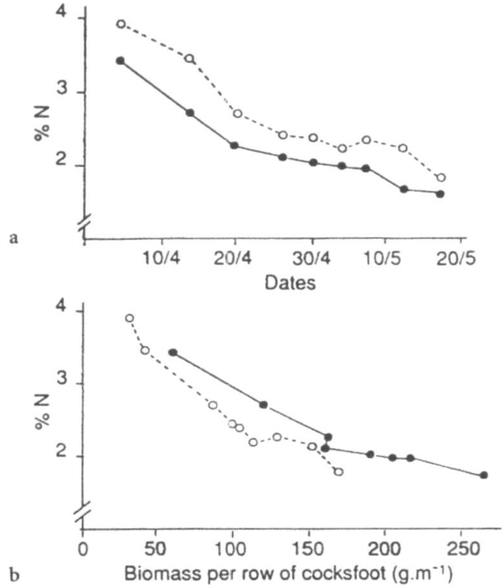

Fig. 8.8a,b. Changes in the N concentration of cooksfoot during spring regrowth in pure stand (●) or in mixture with lucerne (○). **a** Changes with time. **b** Changes with biomass (from Cruz and Lemaire 1986)

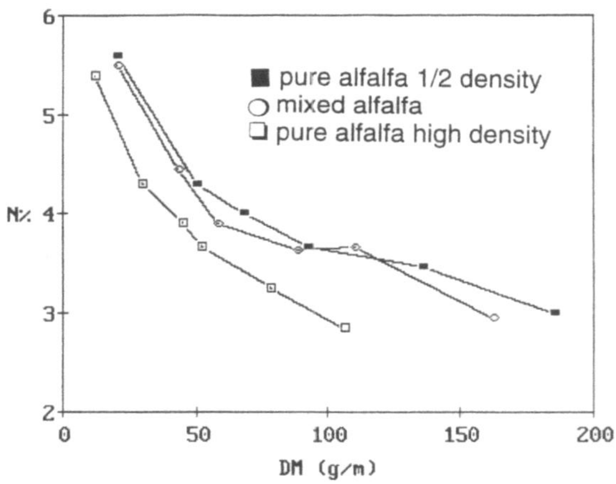

Fig. 8.9. N concentration of *Medicago sativa* in pure stand or mixed to *Dactylis glomerata*, as a function of dry matter accumulation (from Cruz and Lemaire 1985)

This work also allowed an analysis of the N nutrition of the legume species (Cruz and Lemaire 1985). When no fertilizer N was supplied, adding the cocksfoot to the alfalfa crop produced no change in the N nutrition of the legume (Fig. 8.8). This result could also be related to the light relationships between these species: when no N fertilizer is supplied, the grass growth is limited and the legume rows are not shaded. In the same experiment, when alternate rows of a high density alfalfa crop (simple spaced rows) were replaced by cocksfoot, the N nutrition of the legume was improved. This shows the usual substitution effect, i.e., that intraspecific competition is stronger than interspecific competition (Fig. 8.9).

Table 8.1. Legume effect on the nitrogen nutrition of associated grasses, assessed according to the critical N% approach

Design	Species	Growth habit[a]	Growth cycle[b]	N supply	Effect on grass[c]	Reference
Additive	*Dactylis glomerata*	E	P	No	=	Cruz and Lemaire (1986)
	Medicago sativa	E	P	Yes	–	
	Festuca arundinacea	E	P	No	+	Cruz et al. (1991)
	Trifolium repens	S	P	Yes	=	
	Dichanthium aristatum	S	P	No	+	Cruz et al. (1993)
	Gliricidia sepium	E	P			
	Zea mays	E	A	No	–	Clermont-Dauphin (1995)
	Phaseolus vulgaris	E	A	Yes	–	
Substitutive	*Lolium perenne*	E	P	No	+	Soussana and Arregui (1995)
	Trifolium repens	S	P	Yes	+	
	Digitaria decumbens	S	P	No	=	Cruz and Sinoquet (1994)
	Arachis pintoi	S	P	Yes	=	
	Sorghum vulgare	E	A	No	+	Marín et al. (1995)
	Canavalia ensiformis	E	A			

[a] E, Erect; S, spreading.
[b] P, Perennial; A, annual.
[c] =, Neutral; –, negative; +, positive.

8.4 How Legume Species Interfere with Grasses in Mixed Crops

Table 8.1 attempts to summarize the results from experiments where the critical N% approach was used. Such results concern merely the positive, negative, or neutral effect of the legume on the N nutrition of the mixed grass in additive and substitutive designs. It seems clear that in substitutive trials, the grass almost always improved its N nutrition when associated with a legume. However, as was previously pointed out, soil N is in relative terms more available to the grass plants in a mixture due to the reduction of the grass density and the lower competitive ability for N of the legume. On the other hand, additive trials, where the total density of the stand is increased in the mixture, display rather contradictory results, with no clear positive effect of the legume. Yet it is well known that positive effects in legume tree-grass systems are not actually due to the transfer of fixed N, but to the improvement of the water status of the shaded grass (Wilson and Wild 1991).

8.5 Conclusion

Two main conclusions can be drawn from the present study. Firstly, the interference in the N nutrition created by a neighboring species cannot be measured, or even estimated, by the changes in the N concentration of the shoots without taking into account the amount of growth. Secondly, different experimental designs might lead to different conclusions when looking at competitive relationships between species. Each design needs to be well adapted to the hypothesis tested, and the extrapolation of the results should be considered with caution. The planning of the experiment should take into account the density dependence of the shoot N concentration. In the case of mixtures that include spreading species, the N% could be plotted against the best available index of the shading of the species considered. The total radiation intercepted seems to be a good index when the analyzed component is the light-dominant species, but could be inadequate in other cases.

References

Baldy C, Stigter CJ (1993) Agrométéorologie des cultures multiples en régions chaudes. INRA, Paris

Belsky AJ (1992) Effects of trees on nutritional quality of understorey gramineous forage in tropical savannas. Trop Grassl 26:12–20

Clermont-Dauphin C (1995) Relationship between beans and maize intercropped in a ferralitic soil in Haiti. In: Sinoquet H, Cruz P (eds) Ecophysiology of tropical intercropping. INRA, Paris, pp 428–436

Cruz P, Lemaire G (1985) Effet de la compétition sur le niveau de nutrition azotée d'une luzerne en culture pure ou en culture associée à une graminée. In: Guy P (ed) Nutrition azotée des légumineuses. Les colloques de l'INRA no 36, INRA, Paris, pp 257–264

Cruz P, Lemaire G (1986) Analyse des relations de compétition dans une association de luzerne (*Medicago sativa* L.) et de dactyle (*Dactylis glomerata* L.). Effets sur la nutrition azotée des deux espèces. Agronomie 6:735–742
Cruz P, Sinoquet H (1994) Competition for light and nitrogen during a regrowth cycle in a tropical forage mixture. Field Crops Res 36:21–30
Cruz P, Sinoquet H, Gastal F, Moulia B, Varlet-Grancher C, Lemaire G (1991) Efecto del nitrógeno y de la presencia de trébol blanco sobre festuca alta. Turrialba 41:475–481
Cruz P, Munier-Jolain NM, Tournebize R, Sinoquet H (1993) Growth and mineral nutrition in a *Dichanthium aristatum* sward shaded by trees. Proc XVII Int Grassl Congr, New Zealand-Australia, pp 2056–2057
de Wit CT, Tow PG, Ennik GC (1966) Competition between legumes and grasses. Versl Landbouwkd Onderz 687:3–30
Dubach M, Ruselle MP (1994) Forage legume roots and nodules and their role in nitrogen transfer. Agron J 86:259–266
Faurie O (1994) Interactions carbone-azote dans des associations prairiales graminée (*Lolium perenne* L.) légumineuse (*Trifolium repens* L.). Etude d'associations simulées en conditions controlées. Thèse de l'Université Blaise Pascal (Clermont II), Clermont-Ferrand, France, 203 pp
Haynes J (1980) Competitive aspects of the grass-legume association. Adv Agron 33:227–261
Lemaire G, Chartier M (1992) Relationship between growth dynamics and nitrogen uptake for individual sorghum plants growing at different plant densities. Proc 2nd Congr Eur Soc Agron, Warwick University, UK, pp 98–99
Marín D, Cruz P, Sinoquet H (1995) Competition for light and nitrogen in a canavalia-sorghum intercrop. In: Sinoquet H, Cruz P (eds) Ecophysiology of tropical intercropping. INRA, Paris, pp 344–352
Puerto Martin A, Rico Rodriguez M, Gomez Gutierrez JM, Garcia Rodriguez JA, Rodriguez Gonzalez R (1984) Influencia de la encina sobre la composición química del estrato herbáceo. Stud Oecol V:151–168
Russelle MP, Allan DL, Gourley CJP (1994) Direct assessment of symbiotically fixed nitrogen in the rhizosphere of alfalfa. Plant Soil 159:233–243
Sangakkara UR, Attanayake KB (1993) Nitrogen content of *Panicum maximum* when grown with stylo and siratro under two moisture regimes. Proc XVII Int Grassl Congr, New Zealand-Australia, pp 328–329
Soussana JF, Arregui C (1995) Effects de la culture en association sur le niveau de nutrition azotée et la croissance du ray grass anglais et du trèfle blanc. Agronomie 15:81–96
Soussana JF, Arregui C, Hazard L (1989) Assimilation du nitrate et fixation symbiotique dans les associations trèfle blanc-raygrass anglais. Proc XVI IGC, Nice, pp 147–148
Soussana JF, Faurie O, Sinoquet H (1995) Radiation and nitrogen acquisition and use in simulated grass-clover swards. In: Sinoquet H, Cruz P (eds) Ecophysiology of tropical intercropping. INRA, Paris, pp 331–341
Stigter CJ, Baldy C (1995) Manipulation of the microclimate by intercropping: making the best of services rendered. In: Sinoquet H, Cruz P (eds) Ecophysiology of tropical intercropping. INRA, Paris, pp 29–44
Ta TC, Macdowall FDH, Faris MA (1986) Excretion of assimilated N fixed by nodules of alfalfa (*Medicago sativa*). Can J Bot 64:2063–2067
Tournebize R, Sinoquet H (1995) Light interception and partitioning in a shrub/grass mixture. Agric For Meteorol 72:277–294
Tournebize R, Sinoquet H, Bussiere F (1996) Modelling evapotranspiration partitioning in a shrub/grass alley crop. Agric For Meteorol 81:255–272
Vallis I, Haycock KP, Ross PJ, Henzell EF (1967) Isotopic studies on the uptake of soil nitrogen by pasture plants. III. The uptake of small additions of 15N-labelled fertilizer by Rhodes grass and Townsville lucerne. Aust J Agric Res 18:865–877
Vandermeer J (1989) The ecology of intercropping. Cambridge University Press, Cambridge
Virtanen AI, von Hausen S, Laine T (1937) Investigations on the root nodule bacteria of leguminous

plants. XX. Excretion of nitrogen in associated culture of legumes and non-legumes. J Agric Sci 27:584–610
Wilson JR (1990) Agroforestry and soil fertility: the eleventh hypothesis: shade. Agrofor Today 2:14–15
Wilson JR, Wild DWD (1991) Improvement of nitrogen nutrition and grass growth under shadind. In: Shelton HM, Stür WW (eds) Forages for plantation crops. ACIAR Proc No 12, Canberra, pp 77–82

PART III

Management of N Nutrition

CHAPTER 9

Nitrogen Diagnosis and Decision Support

J.M. Meynard[1], C. Aubry[2], E. Justes[3], and M. Le Bail[2]

Contents

9.1 Introduction

The considerable progress made in diagnosing nitrogen nutrition and modeling the impact of nitrogen on growth and yield (see preceding Chaps. 1 to 8) has disclosed many potential applications in crop management: forecasting yields and crop quality, deciding on fertilizer applications when a deficiency is diagnosed, estimating nitrogen requirements according to growth or yield objectives, etc. To develop decision aids, however, one must not only have an understanding of biological processes; one also needs sufficient knowledge of the background to decision making, the decision- maker's constraints and objectives, and even his or her way of thinking (Cerf et al. 1990). As these two spheres are usually considered separately, the aim of this chapter is to attempt a synthesis between them so as to

[1] Laboratoire d'Agronomie INRA INA-PG, 78850 Thiverval-Grignon, France
[2] Unité de Recherche sur les Systèmes Agraires et le Développement INRA INA-PG, 78850 Thiverval-Grignon, France
[3] INRA Unité d'Agronomie, Centre de Recherche Agronomique, 2 Esplanade Roland Garros, BP 224, 51686 Reims Cédex 2, France

Diagnosis of the Nitrogen Status in Crops
G. Lemaire (ed)

define more clearly (1) the qualities required of nitrogen nutrition indicators if they are to be useful in support of decisions, and (2) the procedural basis for developing and evaluating these indicators.

9.2 Nitrogen Diagnosis: an Aid to Which Decisions?

The first stage in developing decision support is to discover how the intended beneficiaries make their decisions, and formalize the process. Two categories of decision-makers are particularly concerned with nitrogen diagnosis: (1) farmers, especially as regards management of nitrogen fertilizer applications, and (2) those responsible for collecting farm produce and supplying the food and agribusiness processing industries, since they must control the quality of batches supplied to factories or put on the market. For a more coherent presentation, we will concentrate on cereals in both cases.

9.2.1 Managing N Fertilizer Applications on the Farm

As a rule, agricultural scientists specializing in nitrogen application merely advise farmers on the optimum choice of dates, quantities, and forms of fertilizer, according to the characteristics of each field (crop species, stage, biomass, production objectives, soil type, past and likely future weather patterns). In this context, nitrogen nutrition indicators can be used to:

1. Estimate a crop stand's nitrogen requirements according to its nutritional status at a given moment, in order to decide on fertilizer rates.
2. Activate the decision to apply fertilizer, if nutrition has recently fallen to a given level, or soon will.
3. Access fertilizer strategy a posteriori: diagnosis of insufficient or excessive nitrogen nutrition late in the season, when all the current season's dressings have already been applied, may lead to rectification for the following year according to the rules adopted for nutrient application planning.

The scientific and technical literature gives many examples of the use of nitrogen nutrition indicators for one or other of these purposes. With cereals, for example, nitrate content in the stem base extract has been used for purposes (1) (Gardner and Jackson 1976; Papastylianou et al. 1984) and (2) (Justes et al., Chap. 10; Wehrmann et al. 1982; Darby et al. 1986; Knowles et al. 1991), while the nitrogen content of the two leaves below the flag leaf (Loué 1984) or NNI (see Chaps. 1 and 2) has been used for a posteriori diagnosis.

However, research into the way in which farmers make their crop management decisions, for nutrient application in particular, adds a great deal to our view of the role of such indicators.

Because the management of a crop involves recurrent decisions[4], and several crops or farm activities must be managed concurrently, farmers plan their technical production management decision to a large extent in advance. Technical decisions on the farm can be represented in the form of forward-planning shedules and sets of decision-making rules (Duru et al. 1988; Sebillotte and Soler 1990). Such forward planning does much to guide farmers in their cultivation operations.

A study of winter wheat management on French farms (Aubry 1995) shows that these planning schedules are designed not field by field, as agricultural scientists implicitly assume, but for all the fields put down to the same crop. The planning schedules concern the timing of cultivation operations (the time of each operation and the order in which they are carried out) and the form they are to take (active molecule, application rate, etc.). This enables us to distinguish several types of decision making rule. Three types of rule determine the timing of operations:

- *Sequencing rules*, which define the chronological order of operations in each field. The order may be necessary (e.g., the previous crop has to be harvested before the wheat is sown) and/or be designed to meet particular crop management objectives (e.g., if a farmer wishes to eliminate all competition from volunteer Gramineae for fertilizer uptake, the operations schedule will put weeding before fertilizer application).
- *Activation rules*, which trigger an operation on a given set of fields when an indicator specific to that operation reaches a certain threshold value. In the JUBIL method, the third nitrogen dressing on a wheat field is triggered when the nitrogen content of the stem base extract falls to a certain threshold value.
- *Arbitration rules* which, at certain times, define priorities among operations when the labor and equipment available on the farm do not allow them to be carried out simultaneously. There are arbitration rules between concurrent crops at a given moment (for example: N fertilization of wheat and sowing of sugar beet are concurrent in the north of France at the end of March and beginning of April).

The form of each operation (and even its timing) is not usually defined field by field. For each technical operation, the farmer defines groups of fields to be treated in the same way; this significantly simplifies the decision-making process. For example, grouping the farm's fields according to the sowing-dates criterion will be used to choose between fields when deciding on the date of the second nitrogen dressing, while grouping according to nature of preceding crop will be used to define nitrogen application rates.

When an indicator measured in situ has to be used to activate a decision-making rule, it has been found that farmers often simplify their work by gathering data on only one field, which Aubry (1995) calls the indicator field. The decision-making process then consists of extrapolating to a group of fields the decision

[4] For example, for annual crops, the farmer knows he must sow, manure, and weed the crop each year on roughly the same dates, according to the average climate of the region.

based on the recorded condition of the indicator field; use of such indicator fields reduces observation time and also simplifies information processing.

With the aid of these few facts on farmers' decision-making processes, we can be more precise about the sphere of application of nitrogen diagnosis. Besides the three uses already identified, that is (1) adjustment of fertilizer rates, (2) triggering of operations, and (3) strategy evaluation, one can imagine the use of indicators in:

4. Arbitration rules.
5. Grouping rules.
6. The choice of indicator fields.

Uses (1) and (2) take place just before fertilizer applications; use (3) takes place at harvest, while uses (4), (5), and (6) come a few weeks before the fertilization. For example, an early nitrogen diagnosis might be carried out on all the wheat fields of a farm, and would then be used for arbitration (e.g., deciding which field needs treatment most urgently), grouping (fields on which the same fertilizer strategy will be suitable) and choice of indicator fields (the only ones that will thereafter be monitored to define fertilizer application dates and quantities).

9.2.2
Quality Management for a Farm Produce-Collecting Area

Grain buyers (the milling and malting industries, exporters, etc.) are diversifying their outlets, innovating their processing methods, raising the standards of their criteria, and trying to base relations with their suppliers on quality assurance. In this way, the suppliers (the grain-collecting and storage firms) now have a full role to play in quality control. Analyzing procedures to facilitate quality control by the grain-collecting firms, Heinz (1992) considers that they manage a veritable industrial process called oriented blending. From the many batches harvested on fields scattered around a more or less heterogeneous area (in terms of environment and cropping systems), the collecting firm makes up a certain number of blended batches that can be delivered to users in line with specifications laid down at the product definition stage.

For soft wheat and barley, the protein content of the grain is a major criterion for quality assessment. It has been shown (Le Bail, research in progress) that control over this blending is very largely the responsibility of the collecting firm, and involves establishing an information system parallel to product flow. In particular, the preharvest information system is designed to give information on:

1. The general state of supply, so that the firm can know whether its preharvest commitments can be met (in quantity and quality) and what other commitments it may be possible to make.
2. The probable quality, in each area, of the batches to be delivered at harvest, since they may have to complete the oriented blending process within the very short period of the harvest.

Classification, now generally based on variety, could be based on protein content, depending on the firm's logistical capability and using different procedures according to the product concerned.

Spring barley for brewing (which in France does not generally involve large quantities) is rarely sorted. A sample is taken on delivery; if the protein content is above 12%, the consignment is paid for at forage barley price, though it is not separated from the malting batch (since the time taken to analyze the sample excludes this possibility). In the long term, it would doubtless be appropriate to set aside high-protein barley batches for livestock feed, to ensure that the malting barley is low in protein. As systematic measurement of protein levels on delivery is not an immediate prospect (would the cost be justified ?), it would be useful to have a procedure for identifying those fields with the highest risk of producing high-protein grain.

For soft wheat varieties that are usable for (1) biscuit-making (requiring a low protein content) and (2) bread-making (requiring a high protein content), it might be useful to introduce a process for separating the wheat into batches with significantly different protein levels, prior to delivery, on the basis of field performance forecasts. This means looking for early protein level indicators that are sufficiently quick and easy to measure to enable a large number of fields to be tested, and the results interpreted, within a short space of time.

Specifications for an early indicator, I, of protein content, P, will differ according to the type of decision-making rule involved:

- If one wishes to set aside the crop from fields where protein levels are greater than a threshold P, for example, one must establish a function F that is stable enough (especially from year to year) to establish the equation P = F(I). The decision-making rule based on early reading of indicators I will then be:

  If, for a given field, I > threshold I [where threshold I = F^{-1} (threshold P)], one must set aside the grain from that field.
- If one wishes to make up batches with significantly different protein levels (as with soft wheat for bread- and biscuit-making), indicator I need only rank crops in increasing order of P. The decision-making rule will be, for example:

  Grain from all fields where I is above the median of value of I for the entire catchment area, is put into one batch; grain from all other fields is put in a second batch.

9.3 Qualities Required in Indicators for Nitrogen Diagnosis

Now that we have a better idea of the wide range of potential uses for nitrogen diagnostic indicators in decision support, we can specify the qualities required of such indicators. Here we shall examine five criteria: specificity, sensitivity, memory, predictive value, and ease of application (Table 9.1).

Table 9.1. Qualities of nitrogen diagnosis indicators, in reference to decision support

Decision	Qualities			
	Specificity	Sensitivity	Memory	Ability to predict
Fertilizer rate	Necessary			Response curve to fertilizer applications
Starting of an operation in the farm; arbitration between operations	Necessary	Necessary		N deficiency in a few days
Definition of groups of fields and of pilot fields	Necessary			Response curve to fertilizer applications
Evaluation of cropping systems and fertilization strategies (a posteriori diagnosis)	Necessary		Necessary	
Oriented blending of harvested products				Quality of harvested products

9.3.1 Specificity

A nitrogen diagnostic indicator is specific if it varies only with nitrogen nutrition. The preceding chapters have shown that the nitrogen content of the aerial parts of the plant is not specific, whereas the nitrogen nutrition index seems to be so.

Specificity is a rare quality. Many crop characteristics are influenced by nitrogen nutrition, and most also depend on carbon nutrition. None of the nitrogen stress indicators suggested by Greenwood (1976) – leaf nitrogen fractions, leaf elongation, leaf area, and carbon dioxide exchange rate – can be regarded as specific. The activity of the enzyme nitrate reductase, suggested as a nitrogen stress indicator by Eilrich and Hagenau (1973) and Verstraeten and Vlassak (1981), is not specific; it depends on light (Sawhney and Naik 1990) and molybdenum deficiency (Blondel and Blanc 1975).

Specificity is essential if an indicator is to be used for decision support (Table 9.1). For a posteriori diagnosis (strategy evaluation), one obviously needs indicators that allow specific evaluation of nitrogen management. Similarly, for adjusting fertilizer rates, activating an operation, or allocating fields to different groups, it is important that the indicators used can be interpreted unambiguously. If one has doubts as to the correct interpretation of an indicator and a decision depends on it, it is not a good indicator. The number of vegetative tillers on wheat, for example, is not a specific indicator: a small number of tillers may mean a nitrogen shortage (in which case a dressing may be called for) or, equally, late development due to late sowing, unfavorable temperatures, etc. (which would justify delaying or reducing fertilizer application).

9.3.2 Sensitivity

The more rapidly a nitrogen nutrition indicator reacts to a change in nitrogen nutrition, the more sensitive it is. Figure 9.1 illustrates the relative sensitivity of several indicators in wheat: nitrate content of leaves, roots, and stem base extract, and total nitrogen content of aerial parts and roots. The most sensitive indicators are the stem base extract and root contents. These enable, by comparing the two experimental treatments, a diagnosis of the onset of deficiency earlier than is possible from the aerial biomass. Justes (1993, and Chap. 10) confirms that stem base extract nitrate is more sensitive than the nitrogen nutrition index (NNI) at the onset of deficiency, but that with marked prolonged deficiency, the NNI remains relatively sensitive, whereas nitrate content is no longer sensitive at all.

Sensitivity is a very important quality for activation or arbitration decisions. In intensive wheat growing, for example, where any nitrogen deficiency is to be avoided, an indicator that is sensitive at the onset of deficiency is a valuable aid for

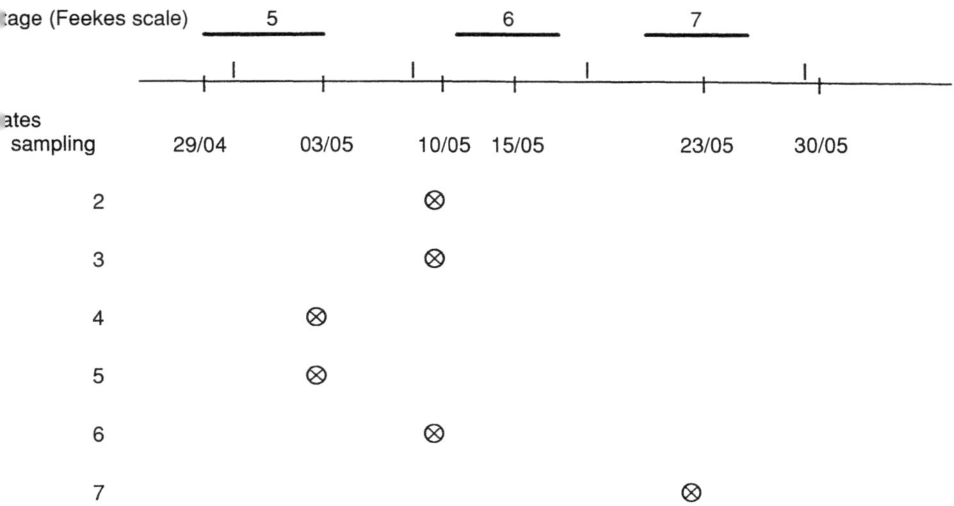

Fig. 9.1. Comparison of the sensitivity of several nitrogen stress indicators at the beginning of N deficiency (Gonzalez-Montaner et al. 1987). ⊗ Date of appearance of significant differences (student's test, significance level: 5%) between an experimental treatment with progressive development of nitrogen stress and a treatment without nitrogen stress, for the aerial biomass, nitrate concentrations and total nitrogen of different organs (Picardie 1985). 2 Aerial biomass. 3 Nitrates in the leaves ($mg\,kg^{-1}$ DM). 4 Nitrate in the roots ($mg\,kg^{-1}$ DM). 5 Nitrates in the "sap" of the stem base ($mg\,l^{-1}$). 6 Total nitrogen content in aerial biomass (%). 7 Total nitrogen content in roots (%)

judging the suitability of additional nitrogen dressings at the stem elongation stage. This is the principle of the JUBIL method (see Chap. 10).

9.3.3 Memory

The memory of an indicator refers to the amount of information it gives about the history of the stand. Yield components, measurable at harvest and reflecting the history of the stand (Sebillotte 1980; Boiffin et al. 1981; Meynard and David 1992) are the classic type of indicator endowed with memory. For cereals, Meynard et al. (1981) have shown that the efficiency of the absorbed nitrogen (yield/nitrogen content of the aerial parts) is a relatively specific global indicator of nitrogen nutrition during the cropping cycle:

N of aerial part: yield < 27 kg ton^{-1} grain $\Rightarrow$ probable nitrogen deficiency.
N of aerial part: yield > 33 kg ton^{-1} grain $\Rightarrow$ probable nitrogen excess.

As 70–80% of the nitrogen in a crop is absorbed before flowering, the yield component number of grains per m^2 calculated before flowering is more strongly affected by variations in nitrogen nutrition than the component thousand-grain weight. Varietal references are needed to interpret the ratio (absorbed nitrogen: grain number), but it allows diagnosis of past deficiencies more specifically than the ratio absorbed nitrogen: yield (Meynard 1987).

Memory is especially necessary for indicators used in fertilizer strategy evaluation indicators, but is not very useful for other aspects of decision support. Sensitive indicators generally have very little memory.

9.3.4 Predictive Value

A predictive indicator makes it possible to infer future elements of crop behavior. Gonzalez-Montaner et al. (1987) consider that a very large nitrate content in the stem base extract (e.g., 6000 mg l^{-1}) enables prediction of an absence of deficiency only for the next 8–10 days. This weak predictive capability apparently explains the failure of fertilizer application methods that use this measurement as an indicator to adjust fertilizer rates. On the other hand, it is sufficient as an activation indicator.

However, although few indicators have a stronger predictive power, they can acquire one if used as input variables in predictive models. For example, the percentage ground cover of the crop canopy during the tillering phase can be used to adjust, on a given field, a model of leaf area index development, on which prediction of the potential growth of the crop and its nitrogen requirements in the absence of any limiting factor can be based (Boissard et al. 1995).

Predictive value is an essential quality for indicators used for crop blending (quality prediction), adjusting fertilizer rates, dividing fields into groups, and choosing indicator fields (prediction of fertilizer response).

9.3.5 Easy Application and Robustness

While the use of indicator fields simplifies in situ information-gathering, the indicators must be easy to monitor: the equipment needed must be inexpensive, must not require specific highly developed skills, etc. The oversensitivity of some indicators may make them impractical as decision aids; for example, farmers are more easily discouraged if variations in an indicator over the course of the day mean that samples must be taken at a specific hour. In any event, in view of the complex organization of labor on a farm, it is very important that decision-making rules be robust, i.e., that they remain pertinent when the indicator is measured at a time other than that recommended, or with sparser sampling, or, more generally, if the rule is applied outside its conditions of strict validity.

9.4 Developing Decision-Making Rules Using Nitrogen Indicators

9.4.1 Designing the Rule

To make a decision, one must project into the future (Sebillotte and Soler 1990), so predictive value is an essential quality for an indicator. As we have seen, this means using the indicator as an input variable in a model predicting quality, occurrence of deficiency, or plant response to fertilizer. Dynamic crop models, the general principles of which are set out by Whisler et al. (1986), are particularly useful because they can be initiated at any moment in the crop cycle for the purposes of prediction. However, as Fisher (1985) emphasizes, these models are very hard to validate for and adjust to a given situation; hence simple, robust, but nondynamic correlative models are often preferred as decision aids (Meynard 1985, 1993). Whatever model is used, knowledge of other input variables – or at least of those to which the model is sensitive – is necessary if the indicator is to be used. Developing a decision-making rule therefore requires, first of all, a thorough examination of the relations between the indicator and the model:

– *Relating the indicator to the variables that make up the crop model*, e.g., biomass, quantity of nitrogen absorbed, leaf area index, or nitrogen nutrition index. Justes (1993, and Chap. 10, this Vol.) analyzes the relationship between stem base extract nitrate content (decision-making indicator) and nitrogen nutrition index (input variable for a simple model predicting the response of yield to nitrogen deficiency). Reau et al. (1994) show that the amount of nitrogen absorbed by oilseed rape at the end of winter (variable needed to predict response of yield to fertilizer) can be estimated from the percentage of ground covered by the crop canopy.

- *Analysis of the model's sensitivity to other input variables.* Those input variables to which the model's outputs (yield, quality, environmental impact, etc.) are sensitive should be taken into account in the decision-making rule. Thorston et al. (1991) specify that variety, soil type, and certain climatic data are always covariables for the decision-making rules they define from their crop process models. Applying JUBIL requires analogous information (wheat variety, growth stage, plant density, soil type, nature of preceding crop, frequency of FYM applications (Justes 1993, quoted in Remy and Hebert 1977). To each value of the nitrogen indicator and its covariables, the decision-making rule links what is regarded as an optimum decision, according to economic calculations based on the model's output variables. Therefore, as is stressed by Rossing et al. (1994) with regard to crop protection, and by Barreto and Bell (1995) for fertilizer application, it is essential to take the uncertainty of prediction into account when defining decision-making rules. In this respect, stochastic models are more appropriate than deterministic models, since one can adapt the indicator's threshold value to the decision-maker's degree of risk aversion.

9.4.2 Evaluating Decision-Making Rules

To evaluate a decision-making rule, one must study the results of its application under farm conditions. One must examine:

- The usefulness of applying it (particularly in comparison with an alternative rule) with regard to different criteria (economic margin, quality of harvested product, environmental impact), which may be aggregated as a usefulness function. The cost of data acquisition for the nitrogen indicator must, of course, be taken into account.
- Its scientific basis: as the validity of a rule depends on the validity of the underlying models and hypotheses, this too must be analyzed. The validation of the JUBIL method is one example. This fertilizer application method involves calculating nitrogen fertilizer rate using the crop balance forecasting model (Remy and Hebert 1977), with an additional rule, which may be formulated as follows:
  - set aside 40 kg N ha^{-1} from the calculated nitrogen quantity X;
  - at key stages, monitor nitrate content of stem base extract (SBE). If, and only if, this falls below a given threshold, apply the 40 kg N ha^{-1} that was set aside.

This rule is based on three separately tested hypotheses:

- H1. If the SBE nitrate content falls below the threshold, the additional application of 40 kg N ha^{-1} is necessary.

- H2. If the SBE nitrate content falls below the threshold, the total amount of fertilizer is as great as if one had simply applied the balance sheet method; one must verify that there is no negative impact on the crop from delaying the 40 kg set aside until the SBE nitrate content has fallen.
- H3. If the nitrate content does not fall below the threshold, an additional N application is not necessary.

To test these three hypotheses, a special system had to be set up, involving three experimental treatments:

- Application of the balance sheet method rule (calculated dose X).
- Application of the JUBIL rule (rate X-40 if nitrate content of SBE does not fall below threshold, rate X if it does).
- Systematic "X-40" treatment to test H1 and H3.

Table 9.2. Results of the test of the three hypotheses underlying the JUBIL method. (Laurent et Justes 1994)

	Yield (dton ha^{-1})		Protein content (%)	
	No. of experiments	Average differences	No. of experiments	Average differences
Hypothesis 1 If N SBE ≤ threshold, an additional application of 40 kg N ha^{-1} is necessary (comparison JUBIL, X-40)	55 V	+5.3*	50 V	+0.9*
Hypothesis 2 If N SBE ≤ threshold, the delay in the application of the additional 40 kg N ha^{-1} has no adverse consequences (comparison JUBIL, balance sheet method X)	58 V	+1.1*	54 V	+0.3*
Hypothesis 3 If N SBE > threshold, no additional application of fertilizer is necessary (comparison X, X-40)	84 V	+0.7	75 NV	−0.5*
Validation of the decision rule (The use of the JUBIL method versus the balance sheet method induces a reduction of the fertilizer rate (average 20 kg ha^{-1}) but no reduction of yield and protein content)	33 V	+0.3	31 V	−0.2

V = Validated hypothesis; NV = Nonvalidated hypothesis.
* Significantly different from 0 ($P = 0.05$).
N SBE = Nitrate concentration in the stem base extract.
JUBIL, X, X-40: see text.

As shown in Table 9.2, all three hypotheses are verified with regards to yield; for protein content, the third hypothesis is not verified, meaning that application of the rule leads to a slight reduction of protein content when the indicator does not fall below the threshold. Following this evaluation, one can characterize the mean advantage and risks associated with the use of the new rule, its sphere of validity (geographical range, cropping system type, values of other indicators, etc., for which the use of the rule is cost-effective) and its robustness (capacity to remain pertinent beyond its strict sphere of validity).

Evaluation of a crop management decision-making rule must be carried out on a multisite experimental system, sampling the range of soils, microclimates, and cropping systems of the region concerned. Such a network, necessarily consisting of on-farm experiments, must be so constructed as to test the hypotheses and models used in constructing the rule. Thus, to test the JUBIL rule referred to above, a trial network was set up throughout France, designed to diversify preceding crop type and earlier FYM inputs. In this way, it was possible to vary the supply of N from the soil (the amount of nitrogen absorbed in the absence of fertilizer varied in this network from 70 to 320 $kg\,N\,ha^{-1}$) and, above all, the uncertainty of prediction: the greater the uncertainty, the more it may be considered helpful to use a plant nutrition indicator to decide on fertilizer applications. By studying several varieties, it was possible to analyze the robustness of the intervention threshold according to genotype.

Evaluation of a rule designed for organizing collection does not require specific experimentation, but can be done by simulating the blending of batches from a network of farmers' fields. Table 9.3 gives an example of evaluation of two rules for batch blending for brewing barley, where the aim is batches with a fairly low average protein content. The first rule is to define the destination of the crop (brewing or rejected batch) by rejecting those 25% of fields that have the highest content in the two leaves below the flag leaf (Loue's foliar diagnosis, 1984). The second rule, more easy to apply and apparently more effective, consists of rejecting all lodged fields.

Table 9.3. Evaluation of two decision rules for oriented blending in malting barley. (Le Bail, unpubl.)

Decision rule	No selection	Selection on the basis of the N content in upper leaves		Selection on the basis of observation of lodging	
Blend	Blend of the whole harvested grain	Malting blend (75% of the plots)	Refused blend (25% of the plots)	Malting blend (plots without lodging)	Refused blend (plots with lodging)
No. of plots	35	27	8	25	10
Average protein content of the blend (%)	9.7	9.3	11.4	9.2	11.3

9.5 Conclusion

We have tried to show that the present state of the art allows a scientific approach to developing diagnostic indicators and decision-making rules. While diagnosis, prediction, and operational guidance depend on knowledge of internal crop processes, they cannot be reduced to that. Indicators and rules must not be regarded as research applications but as fully fledged objects of research.

At present, the range of indicators that scientists can offer decision-makers is fairly small. With increasing understanding of nitrogen nutrition and advances in measurement techniques, it should be possible to offer more indicators within the next few years. Given the diversity of farmers' needs, farm structures, and organization of work, it will be useful to have a wide range of indicators and decision-making rules. Our growing knowledge of the background to decision-making opens up some fascinating lines of research: agricultural scientists have so far worked primarily on indicators for adjusting fertilizer rates and activating application; they have not yet gained a thorough knowledge of the qualities required of indicators for field grouping, a posteriori diagnosis, or batch blending.

Developing decision-making rules should also, increasingly, involve crop process models, which alone can guarantee improvements to the predictive value of the indicators. Evaluation, a necessary step prior to widespread extension work, is a new kind of experimental activity. It differs both from conventional experiments designed to determine optimum techniques (variety trials, fertilizer response curves) and from demonstration trials. The development of indicators and the associated decision-making rules is thus becoming a scientific field in itself, at the crossroads between agronomy, plant physiology, and decision analysis.

References

Aubry C (1995) Gestion de la sole d'une culture dans l'exploitation agricole. Cas du blé d'hiver en grande culture dans la région picarde. Thèse INA-PG, Paris, 279 pp + Annexes

Barreto MJ, Bell MA (1995) Assessing risk associated with N-fertilizer recommendations in the absence of soil tests. Fertil Res 40:175–183

Blondel AM, Blanc D (1975) Accidents végétatifs sur agrumes en Corse. Mise en évidence d'une carence en molybdène par utilisation du test de mesure in vivo de l'activité de la nitrate réductase. Ann Agron 26:277–287

Boiffin J, Caneill J, Meynard JM, Sebillotte M (1981) Élaboration du rendement et fertilisation azotée du blé d'hiver en Champagne crayeuse. I. Protocole et méthode d'étude d'un problème technique régional. Agronomie 1:549–558

Boissard P, Akkal N, Valéry P, Lewis P, Meynard JM (1995) 3D-Plant characterization and modelling aimed at the remote control of winter wheat growth. Communication au Colloque Photosynthèse et télédétection, août 1995, Montpellier

Cerf M, Papy F, Aubry C, Meynard JM (1990) Théorie agronomique et aide à la décision. In: Brossier J, Vissac B, Le Moigne JL (eds) Modélisation systémique et système agraire. Décision et organisation. INRA Paris, pp 181–202

Darby RJ, Widdowson FV, Bird E, Hewitt MV (1986) The relationship of soil mineral NO_3^-N with stem NO_3^-N concentration, and of fertilizer-N with the amount of nitrogen taken up by winter wheat, in experiments testing nitrogen fertilizer in combination with aphicides and fungicides, from 1980 to 1982. J Agric Sci Camb 106:497–507
Duru M, Papy F, Soler LG (1988) Le concept de modèle général et l'analyse du fonctionnement de l'exploitation agricole. CR Acad Agric Fr 74:81–93
Eilrich LG, Hageman RH (1973) Nitrate reductase activity and its relationship to accumulation of vegetative and grain nitrogen in wheat. Crop Sci 13:59–66
Fisher RA (1985) The role of crop simulation models in wheat agronomy. In: Day W, Atkin RH (eds) Wheat growth and modelling. Plenum Press, New York, pp 237–255
Gardner BR, Jackson EB (1976) Fertilization, nutrient composition and yield relationships in irrigated spring wheat. Agron J 68:75–78
Gonzalez-Montaner JH, Meynard JM, Mary B (1987) Contrôle de la nutrition azotée du blé par l'analyse des teneurs en nitrate dans la plante. CR Acad Agric Fr 73:105–115
Greenwood EAN (1976) Nitrogen stress in plants. Adv Agron 28:1–35
Heintz W (1992) Modèles d'entreprise et conventions de qualité. Les entreprises de collecte-stockage sur les marchés du blé. Thèse INA-PG, Paris, 289 pp + Annexes
Justes E (1993) Diagnostic de la nutrition azotée du blé à partir de la teneur en nitrate de la base de to tige. Application au raisonnement de la fertilisation. Thèse INA-PG, Paris, 227 pp + Annexes
Knowles TC, Doerge TA, Ottman MJ (1991) Improved nitrogen management in irrigated durum wheat using stem nitrate analysis. II. Interpretation of nitrate-nitrogen concentrations. Agron J 83:353–356
Laurent F, Justes E (1994) Fertilisation azotée du blé d'hiver: JUBIL, un nouvel outil de pilotage. Perspect Agric 190:62–69
Le Bail M (1994) Crop quality management: research questions. In: Borin M, Sattin M (eds) Proc 3rd ESA Congr Abano-Padova, pp 560–569
Loué A (1984) Blé-orge. In: Martin-Prével P, Gagnard J, Gautier P (eds) L'analyse végétale dans le contrôle de l'alimentation des plantes tempérées et tropicales. Lavoisier, Paris, pp 643–674
Meynard JM (1985) Construction d'itinéraires techniques pour la conduite du blé d'hiver. Thèse INA-PG, Paris, 297 pp
Meynard JM (1987) L'analyse de l'élaboration du rendement sur les essais de fertilisation azotée. Perspect Agric 115:76–83
Meynard JM (1993) The use of crop models to introduce new management systems. In: Peeters A, van Bol V (eds) Potential and limits of organic farming. Commission of the EEC, Université Catholigue de Lourain-la-Neuve, Belgium, pp 25–41
Meynard JM, David G (1992) Diagnostic sur l'élaboration du rendement des cultures. Cah Agric 1:9–19
Meynard JM, Boiffin J, Caneill J, Sebillotte M (1981) Elaboration du rendement et fertilisation azotée du blé d'hiver en Champagne Crayeuse. II. Types de réponse à la fertilisation azotée et application du bilan prévisionnel. Agronomie 1:795–806
Papastylianou I, Graham RD, Puckridge DW (1984) Diagnosis of the nitrogen status of wheat at tillering and prognosis for maximal grain yield. Commun Soil Sci Plant Anal 15:1423–1436
Reau R, Paillard C, Wagner D, Jung I (1994) Un outil pour raisonner la fertilisation azotée du colza. Oléoscope 19:22–23
Rémy JC, Hébert J (1977) Le devenir des engrais azotés dans le sol. CR Acad Agric Fr 63:700–710
Rossing WAH, Daamen RA, Jansen MJW (1994) Uncertainty analysis applied control of aphids and brown rust in winter wheat. 2. Relative importance of different components of uncertainty. Agric Syst 44:449–460
Sawhney SK, Naik MS (1990) Role of light in nitrate assimilation in higher plants. In: Abrol YP (ed) Nitrogen in higher plants. Research Studies Press, Somerset, pp 93–128
Sebillotte M (1980) Analyse de l'élaboration du rendement du blé: un moyen pour résoudre les problèmes de cette culture. In: Wheat: technical monograph. CIBA-GEIGY, Bâle, pp 25–32
Sebillotte M, Soler LG (1990) Les processus de décision des agriculteurs. In: Brossier J, Vissac B, Le Moigne JC (eds) Modélisation systémique et système agraire: décision et organisation. INRA, Paris, pp 93–117

Thorston RK, Dent JB, Bacsi Z (1991) A framework for crop growth simulation model applications. Agric Syst 37:327–340

Verstraeten LMS, Vlassak K (1981) Nitrogen-stress and plant growth in relation to the nitrogen status of plant and soil. Pédologie 31:379–392

Wehrmann J, Sharpf HC, Böhmer M, Wolbring J, (1982) Determination of nitrogen fertilizer requirements by nitrate analysis of the soil and of the plant. 9th Int Plant Nutrition Coll Warwick Univ, Warwick

Whisler FD, Acock B, Baker ON, Fye RE, Hodges HF, Lambert JR, Lemmon NE, McKinion JM, Reddy VR (1986) Crop simulation models in agronomic systems. Adv Agron 40:141–208

CHAPTER 10

Diagnosis Using Stem Base Extract: JUBIL Method

E. JUSTES[1], J.M. MEYNARD[2], B. MARY,[3] and D. PLÉNET[4]

Contents

[1]INRA Unité d'Agronomie Centre de Recherche Agronomique, 2 Esplanade Roland Garros, BP 224, 51686 Reims Cédex 2, France
[2]INRA-INA PG Laboratoire d'Agronomie, 78850 Thiverval-Grignon, France
[3]INRA Station d'Agronomie, Rue Fernand Christ, 02007 Laon Cédex, France
[4]INRA Unité d'Agronomie, 71, av. Edouard Bourleaux, BP 81, 33883 Villenave d'Ornon Cédex, France

Diagnosis of the Nitrogen Status in Crops
G. Lemaire (ed)

10.1 Introduction

In this chapter, the principle and validation of the JUBIL method are presented. The value and limits of this method for winter wheat and maize crops are then analyzed. This research was carried out in collaboration with ITCF[5] and AGPM[6].

10.2 Methods of N Fertilization Recommendations

10.2.1 Use of a Soil Indicator: the Balance-Sheet Method

In order to reduce losses of nitrogen (N) from agricultural areas to atmosphere and water, N fertilization of arable crops must be monitored as precisely as possible. In France, N fertilization recommendations for the major crops are mainly based on the balance-sheet method (Hébert 1969; Rémy and Hébert 1977; Rémy 1981). The balance considers variations in the soil mineral N in the potential rooting zone between midwinter and crop harvest. A predictive balance is made to determine the amount of fertilizer N needed to supplement the soil N supply to meet the crop N requirements of nonlegume crops. The most complete balance of the mineral N pool can be written (in $kg\,N\,ha^{-1}$):

$$R_f - R_i = (M_n + F + A) - (N_f - N_i + G + L),$$

where R_f = residual inorganic N in the soil at harvest[7]; R_i = residual inorganic N in the soil at mid-February[5]; M_n = net mineralization in soil; F = fertilizer N; A = atmospheric N (dry and wet) deposition and nonsymbiotic fixation; N_f = crop N accumulation at harvest; N_i = crop N accumulation at mid-February[8]; G = gaseous losses (ammonia volatilization and denitrification); L = leaching losses[9].

The simplification which is usually made is that the atmospheric inputs are compensated for by the gaseous losses (A ≅1 G, assessed at about 15–25 kg N ha^{-1} in arable cropping systems; Rémy 1981). In the case of cereals, the N crop uptake can be considered as proportional to the yield:

$$N_f = bY,$$

where bY = crop N requirements, b = amount needed per unit of grain (30 kg N t^{-1} for winter wheat; Coïc 1956) and Y = grain yield target (t ha^{-1}). Net mineralization (M_n) is also described as the sum of three contributions:

[5] ITCF: Institut Technique des Céréales et des Fourrages (France)
[6] AGPM: Association Générale des Producteurs de Maïs (France)
[7] A depth z greater or equal to the rooting depth
[8] This may be large for early sown winter crops (rape, barley)
[9] Occurring below depth z between mid February (R_i) and harvest (R_f)

$$M_n = M_h + M_r + M_a,$$

where M_h = net mineralization from humus, M_r = net mineralization from residues of previous crop, and M_a = net mineralization from organic (fresh and old) amendments.

The simplified balance-sheet model (Machet et al. 1990) can therefore be written:

$$bY + R_f = (R_i - L) + (M_h + M_r + M_a) + F.$$

The different components constituting the soil N supply are either measured (R_i), or estimated from models (M_h and L) or references (b, M_r, and M_a). The N fertilizer requirements of various arable crops (wheat, barley, maize, sugarbeet, potato, rapeseed, vegetables, etc.) can be evaluated for each individual field using the updated computer program AZOBIL (Machet et al. 1990).

This method is based on the time course of inorganic soil N, and a soil indicator is used to determine the initial N content. The soil indicator corresponds to the measurement (over the rooting depth) of inorganic N (NO_3^- and NH_4^+) present in the soil at mid-February (R_i). The balance-sheet method is well adapted to loamy and deep soils for various crops (Machet et al. 1990), in situations where little water deficit occurs, and has been used successfully in chalky soils for winter wheat (Meynard et al. 1981). It has been shown more recently that the use of the balance-sheet method minimizes the risks of nitrate pollution without penalizing yield (J.M. Machet, pers. comm.).

However, since this method gives a prediction over a long period of time (from mid-February to harvest), it may not be reliable enough if the soil N supply and root interception are not correctly estimated. The risk of poor prediction is greater in situations where the soil N supply is large: this is particularly the case for summer crops, after legumes or grasses, and after a fallow or applications of organic wastes rich in N. Under these conditions, the fertilizer N recommendations are small or zero, and farmers often apply larger rates to ensure that no N deficiency will occur. Consequently, the nitrate pollution risks are increased. These situations highlight the interest of using another indicator that allows more precise adjustment of the N fertilizer rate to meet plant requirements.

10.2.2
Use of a Plant Indicator: the Nitrate Concentration in Stem Base Extract (SBE)

It has been shown previously (Chaps. 1 and 2, this Vol.) that the nitrogen nutrition index (NNI) allows a reliable crop N diagnosis. Unfortunately, the determination of NNI requires a measurement of both the aerial biomass and its N concentration, which cannot be routinely done by farmers. It is therefore necessary to propose more practical plant tests.

Various authors (e.g., Papastylianou et al. 1982; Wehrmann et al. 1982, Scaife and Stevens 1983; Darby et al. 1986; Gonzalez-Montaner 1987; Gonzalez-Montaner et al. 1987) have pointed out the interest of measuring plant nitrate concentration to evaluate the N nutrition status of vegetables or cereals (nitrate test). The nitrate concentration in the stem base extract has been particularly studied for the winter wheat crop because of its sensitivity and suitability for making rapid tests (semi-quantitative measurement of nitrate or the snappy sap test). The nitrate concentration of the stem base extract has been shown to be of interest for diagnosis of the N nutrition status of wheat (Papastylianou et al. 1982; Gonzalez-Montaner 1987). However, this nitrate test cannot be used alone to predict the development of the N crop status on a long-term basis (for example, measurements at tillering are inappropriate to predict N status up to flowering).

Gardner and Jackson (1976) and Papastylianou et al. (1984) proposed the use of measurements made on the stem dry matter for long-term predictions, but other authors (Beringer and Hess 1979; Pettygrove et al. 1984) concluded that it was impossible to establish a useful nitrate concentration threshold to make a long-term diagnosis. More recently, it has been shown that the long-term predictive critical concentration of maize crops can vary according to growth stage (Geyer and Marschner 1990), climatic conditions (Iversen et al. 1985), soil type (MacClenahan and Killorn 1988; Binford et al. 1990), and light intensity (Fox et al. 1989). This is expected, considering that net N mineralization, at least, fluctuates with climatic conditions and soil type.

On the other hand, it appears that the nitrate test can be used to predict the development of N crop status in the short term, from tillering to the onset of stem elongation (Darby et al. 1986; Gonzalez-Montaner 1987) or from the booting stage to ear emergence (Wehrmann et al. 1982; Knowles et al. 1991); but the nitrate test does not allow determination of the amount of fertilizer N required to remove any limitation of growth by N availability from the soil. It is necessary to build a new N fertilizer strategy that integrates a plant indicator with the previously N fertilizer application scheme.

10.2.3 Use of a Combination of a Plant and Soil Indicator: the JUBIL Method

We have suggested the combination of the balance-sheet method (using a soil indicator) with the nitrate test method (using a plant indicator): both indicators are used to form decision-making rules which define the JUBIL method (Justes 1993; Justes et al. 1994a).

In situations where the balance-sheet method gives an unsatisfactory prediction of fertilizer-N amount, it is likely that it will overestimate the amount to be applied. We propose a reduction in the balance-sheet N fertilizer dose (corresponding to a higher soil N supply), to avoid nitrogen excess (a cause of nitrate pollution). Then, we need to monitor the N crop status during growth to detect situations where it is necessary to add N fertilizer to compensate for insufficient soil N supply. The nitrate test is used during stem elongation (Feekes 6 to 9; Feekes scale from Large 1954) to determine whether a supplementary N dressing is needed. Its use corresponds to a short-term predictive N diagnosis.

At the present time, the JUBIL method has been developed for winter wheat cereals. It involves four successive steps:

1. Calculate the overall N fertilizer requirements using the balance-sheet method: this defines the total rate, X (kg ha^{-1});
2. Apply a reduced amount of fertilizer N (X minus 40 kg ha^{-1}), considering that favorable conditions of crop N uptake may happen. One part (40–60 kg ha^{-1}) is applied at tillering (Feekes 3) and the remainder at stage ear 1 cm (stage Feekes 5);
3. Measure the NO_3^- concentration in the main stem base extract during stem elongation to detect N deficiency and to predict the short-term N status of the crop: one to three measurements are needed, depending on the situation. The first measurement is made at stage Feekes 6 (first node). The second measurement takes place at Feekes 7 (second node), and the third at Feekes 8–9 (flag leaf emergence).
4. Apply or omit the last N dressing (40, 60 or 80 kg N ha^{-1}), according to the NO_3^- concentration of the stem base and the growth stage. An N application is required to maintain the target yield if NO_3^- concentration falls below a threshold value.

10.3 Measurement, Interpretation, and Calibration of the Nitrate Test

10.3.1 Measurement Protocol

The nitrate test is carried out on fresh plant material. A 2-cm-long segment is cut at the base of the main stem. This is mainly tissue without chlorophyll. The extraction is made with a small manual press (Routchencko 1967; Gonzalez-Montaner 1987). Around 60 main stem bases are collected together to represent a farmer's field. Results are expressed in mg $NO_3 l^{-1}$ of stem base extract. The sampling is carried out in the field, between sunrise and 2 h after. During transport, the stems must be protected against light and excessive temperature (preferentially in an ice box). On arrival in the laboratory (or at the farm), the roots are cut at the tillering plate level (i.e., at the crown); the dead leaves, leaf sheaths, and diseased stems are removed; the stems are washed and wiped; the stem bases are cut and then pressed; and the extract is diluted with deionized water (1:10) to avoid coloration interference when using rapid test strips. Under laboratory conditions, the nitrate concentration is measured precisely using the Griess method. Under farm conditions, the nitrate concentration is measured with rapid test strips (e.g., Merckoquant or Reflectoquant nitrate), which can be read with a hand reflectometer (e.g., Nitrachek or RQflex). The reflectometer reading yields the nitrate concentration with less than 10% error, which is good enough for making the N diagnosis.

The extract is sometimes erroneously called sap; we prefer to call it stem base extract (SBE), since it is a mixture of different solutions: apoplasmic water, xylem sap, phloem sap, cytosollic and vacuolar water, the vacuolar liquid being the major contributor of SBE. The nitrate concentration of the SBE will be referred to in the following as NSBE.

10.3.2
Interest and Physiological Significance of NSBE

First of all, the nitrate test (NSBE) can be used if the absorption form of N is primarily nitrate and not ammonium. In that case (usual in France and under temperate climate), the NSBE must fulfill several conditions in order to be useful in the diagnosis of plant N status:

1. Its variation must be small during the day/night period.
2. Its change must be large when the crop becomes deficient in N.
3. The activity of nitrate reduction in the roots must be small, so that the major N component transported into the xylem sap is nitrate (and not amino acids).
4. The indicator must be a short-term integrator of the N absorption flux (i.e., over a few days). It must be neither an instantaneous signal such as the xylem nitrate flux nor a signal that is an integral over several weeks, such as the nitrogen index (NNI).
5. The relationship between NSBE and NNI must be significant; so that it is possible to determine the N fertilizer decision-making threshold value of NSBE.

These five conditions are discussed in turn below.

10.3.2.1
Time Course of NSBE During the Day

The variation of NSBE during the day/night period has been studied in winter wheat crops by Justes (1993). An example of NSBE evolution for three N treatments is shown in Fig. 10.1; the measurements were made at Feekes 7 stage. In this example, the NNI indicated that treatment N_1 was deficient, treatment N_2 was starting to lack N, and N_3 was N-deficient, although it had received an N dressing 3 days before the measurements. The NSBE tended to decline during the day in treatments N_1 and N_2, whereas the trend was for an increase in N_3. The latter could result from a dominant effect of the recent N dressing over the daily effect, this daily effect being a slight decline during the day and an increase during the night. However, in such cases, the spatial variability of the nitrate concentration between experimental blocks is often equal to or greater than the daily variation, as is the case in Fig. 10.1.

In order to avoid this possible daily variation, we have recommended that the SBE samplings should be made at the same period of the day, the 2-h period following sunrise being preferred (Justes 1993). However, this is a precaution rather than a necessity.

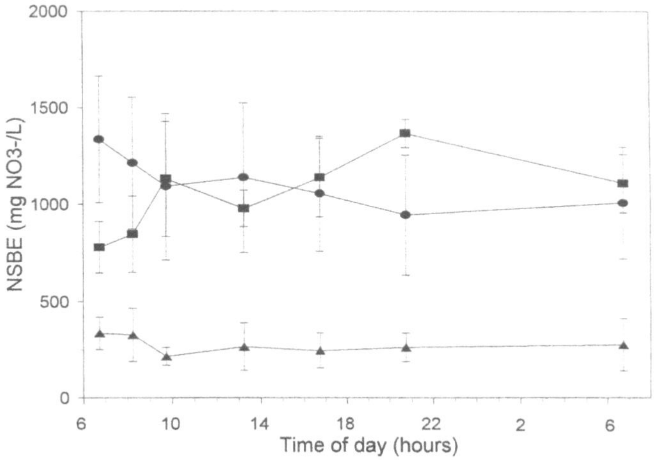

Fig. 10.1. NSBE versus time of day for winter wheat crops at Feekes 7 stage (experiment at Grignon, France, 1992). ▲ N1 treatment; ● N2 treatment; ■ N3 treatment (see text); *vertical bars* standard error

10.3.2.2
Time Course of NSBE During the Crop Growth Cycle

Figure 10.2a shows an example of the time course of the NSBE during crop growth in one experiment including five N treatments (15, 65, 100, 130, and 180 kg N ha^{-1}), taken from Justes (1993). In Fig. 10.2b, which shows the plant N% curves for the five treatments and the critical plant N% curve (Justes et al. 1994b), N180 treatment is N-nonlimiting during all the period monitored (tillering to anthesis), but the other treatments are N-limiting at different growth stages. The NSBE quickly decreases when N deficiency approaches and becomes very low (near zero) when N deficiency is evident; this is the case for the four low N treatments. The NSBE also decreased in the nonlimiting N treatment (N180), but the rate of decline was much slower than in the N-limiting treatments. This is generally the case in experiments carried out in France, except in situations where the soil N supplies are very large (cropping systems with large applications of manure), while the SBE nitrate concentration can remain high (or slightly decrease) during stem elongation, and then decreases after appearance of the ears.

Comparison of the time course of NSBE and NNI in two out of the five treatments previously studied (Fig. 10.2c) shows that these two indicators reacted at different speeds. The NSBE decreased after the appearance of N deficiency and did not increase later when N deficiency was evident. The NNI did not change, or decreased slightly but not significantly. Later, NNI decreased regularly even if the NSBE did not alter.

These two indicators are complementary tools for diagnosis: the NBSE is the indicator for earliest detection of the appearance of N deficiency (Gonzalez-Montaner 1987; Justes 1993), while NNI allows us to quantify it (Justes 1992).

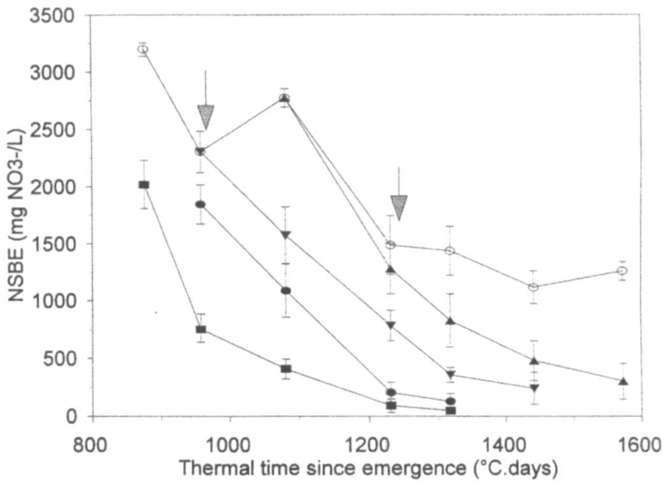

a

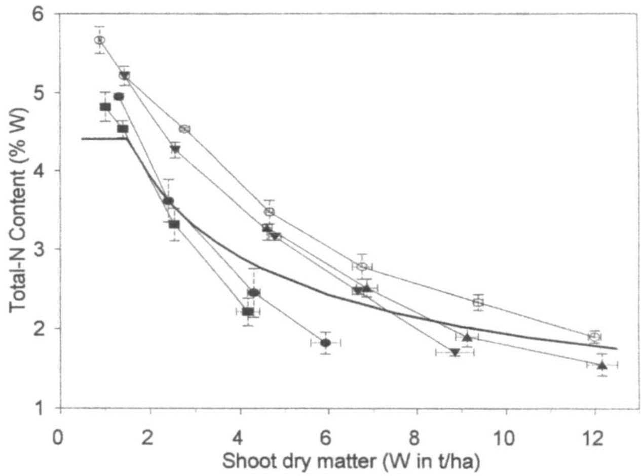

b

Fig. 10.2. **a** NSBE versus thermal time from emergence for winter wheat crops for five N treatments. **b** Plant N% curves observed for the five N treatments and critical N dilution curve ($N_{ct} = 5.35W^{-0.442}$). ■ N15 treatment (which received $15\,kg\,N\,ha^{-1}$); ● N65 treatment; ▼ N100 treatment; ▲ N130 treatment; ○ N180 treatment (N nonlimiting). **c** NSBE (primary y-axis) and NNI (secondary y-axis) versus thermal time from emergence for two N treatments: N180 (N nonlimiting) and N15 (N limiting since Feekes 6); □ NSBE and ■ NNI for N15 treatment; ○ NSBE and ● NNI for N180 treatment; *vertical* and *horizontal bars* represent standard error; *arrows* indicate fertilizer N application for N180 treatment (experiment at Grignon, France, spring 1991, from Justes 1993)

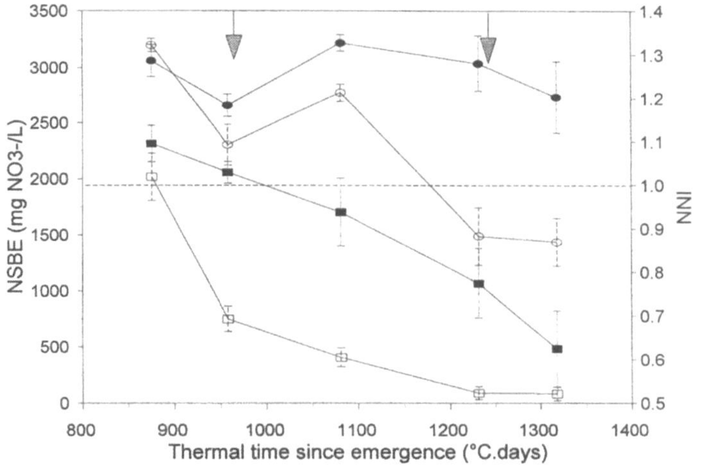

Fig. 10.2c

10.3.2.3
Time Course of NBSE After Application of Various N Fertilizers

Nitrate is known to be the major form of N absorption of arable crops (except rice) in temperate climatic conditions (e.g., Beevers and Hageman 1980), because nitrification is a rapid phenomenon in normal soil conditions (e.g., Recous et al. 1988). After the application of an N fertilizer containing ammonia, plants can take up a large proportion of their N as ammonia over a short period; but across the growing period as a whole, arable crops mainly take up nitrate. However, Justes (1993) showed that after an N application during stem elongation (Feekes 6 to 10), the nitrogen index, NNI, always increased, but the NSBE reacted differently according to the chemical form of the fertilizer and the soil moisture. Under these experimental conditions, the NSBE increased when the fertilizer form was NH_4NO_3 or Ca $(NO_3)_2$ (Fig. 10.3a), but did not increase when urea was used (Fig. 10.3b). N fertilizer can be taken up by the foliar route; in this case, nitrate is immediately reduced to amino acids by the leaves, and so the nitrate cannot accumulate in the stem. Thus, N diagnosis based on the NSBE indicator can be wrong (Justes 1993).

Therefore, no NBSE measurement should be made to evaluate the N status of a winter wheat crop when the deficiency has become large and after a third N application during stem elongation.

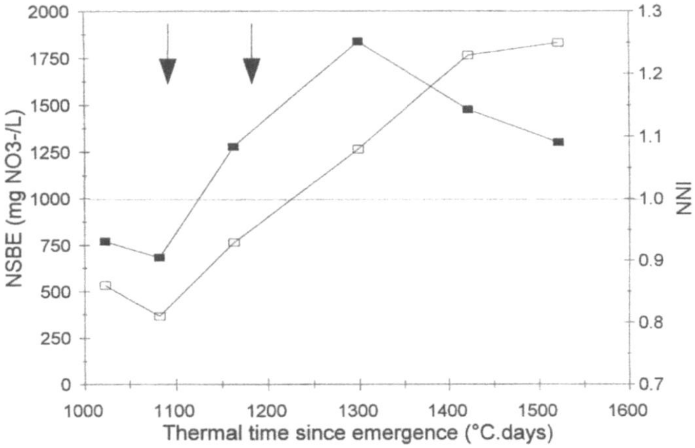

a

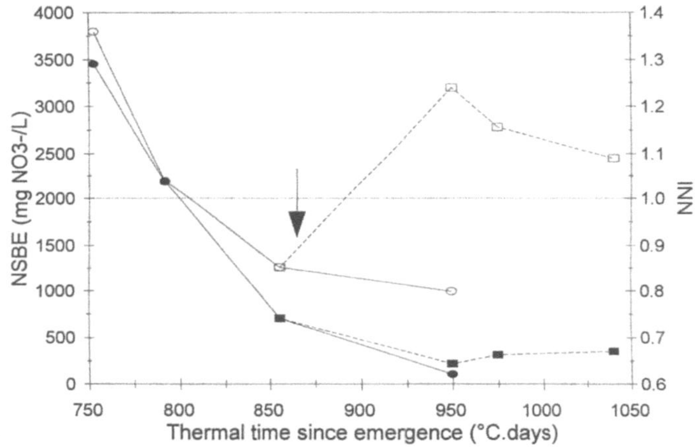

b

Fig. 10.3a,b. Curves of NSBE and NNI versus thermal time from emergence for winter wheat crops after fertilizer N application. (Justes 1993). **a** Fertilizer N corresponding to NO_3NH_4 application on two occasions with $50\,kg\,N\,ha^{-1}$ (Laon 1990): ■ NSBE; □ NNI. **b** Fertilizer N corresponding to urea application with $150\,kg\,N\,ha^{-1}$ (Grignon 1991). ● NSBE; ○ NNI without fertilizer N application; ■ NSBE; □ NNI after fertilizer N application

10.3.2.4
The Contribution of Roots to Nitrate Reduction in Winter Wheat Crops

Nitrate reduction, which is the major source for N nutrition of higher plants, is believed to occur principally in green leaves (Beevers and Hageman 1980). Although numerous results in the literature support this hypothesis, some authors have proposed that roots may have a significant role in reduction, particularly during early growth (Talouzie et al. 1984; Gojon et al. 1986). However, most of the studies were carried out with young plants in the greenhouse or growth chamber.

Furthermore, the methodology of estimating root nitrate reduction using either NRA (nitrate reductase activity) measurements or the ratio of amino N to total N (reduced N plus nitrate N) in the xylem sap has been contested, because of the rapid translocation of amino acids between shoots to roots via the phloem and their return to roots via the xylem (Cooper et al. 1986a). The best methodology relies on the determination of the ratio of amino-^{15}N to total-^{15}N over short intervals after $^{15}NO_3^-$ application. This methodology has been applied to field experiments at different growth stages by Cooper et al. (1986b) and Justes (1993). They both show that, for winter wheat crops during stem elongation, less than 30% of the nitrate is reduced in the roots, whatever the growth conditions. Furthermore, Justes (1993) reported a close relationship between the contribution of roots to nitrate reduction and the N absorption rate measured daily: the root contribution increased when N absorption increased, a result opposite to that of Beevers and Hageman (1980) for young plants of barley. Thus, during tillering and stem elongation (Feekes 3 to 10.5), with an N absorption rate of about 2.5 kg N ha^{-1} in northern France, the contribution of the root is on average around 10–15% of the total nitrate reduction. Consequently, the major part of nitrate absorbed by the roots during the growth cycle is reduced in the shoot.

Hence, we can conclude that an aerial tissue such as the stem base extract could be used to reflect N absorption of the whole winter wheat.

10.3.2.5
Relationship Between Nitrate Concentrations in SBE and Xylem Sap

Justes (1993) showed that the SBE nitrate concentration was 1.5 to 7 times higher than that of the xylem sap, according to growth stage and the time of day that measurements were made. He carried out a field experiment with $^{15}NO_3^-$ (labeled at 50%) to assess the turnover time of nitrate in the stem base and in the xylem sap. The ^{15}N atom% excess of nitrate in xylem sap and SBE was measured between 3 and 72 h following the $^{15}NO_3^-$ application, for various developmental stages. Figure 10.4 shows the results obtained at Feekes 7 stage. The $^{15}NO_3^-$ applied was rapidly absorbed and translocated to shoots via the xylem sap. The isotopic excess of the xylem sap increased up to 48 h after application and then decreased. The same pattern was observed for the $^{15}N\text{-}NO_3^-$ atom% excess of the SBE, the latter being much less labeled. The ratio of the two isotopic excesses (xylem sap:SBE) decreased from 24 to 5. This indicates that the nitrate in the SBE became progressively equilibrated with the nitrate in the xylem sap. This equilibration is, however, not complete 3 days after ^{15}N application. Equilibrium was obtained after 2 days at stage Feekes 7 (ratio = 1). These differences between stages may be explained by the volume differences in xylem sap and SBE according to growth stages.

The nitrate test does not reflect the nitrate concentration of the xylem sap. Thus, daily variations in transpiration do not significantly influence directly the NSBE, which is why the daytime variation of the latter is relatively limited.

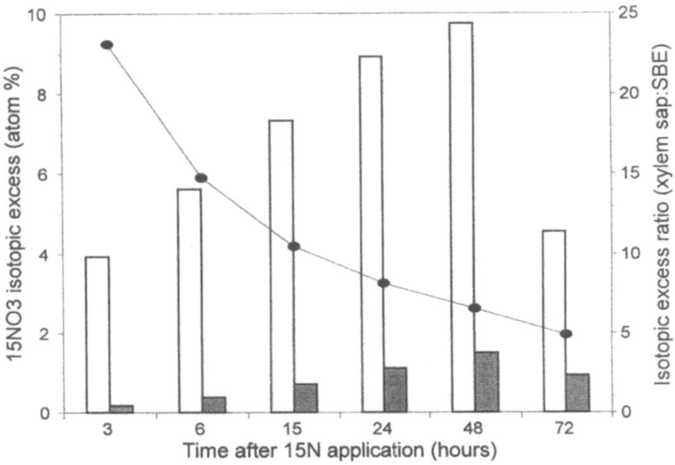

Fig. 10.4. Nitrate isotopic excess (^{15}N%) in the xylem sap (□) and in SBE (■) versus hours after ^{15}N application (primary y-axis) for winter wheat crop. Isotopic ^{15}N excess ratio between xylem sap and SBE (-●-) versus hours after ^{15}N application (secondary y-axis) (experiment at Grignon, France, April 1992, at Feekes 7 stage, from Justes 1993)

10.3.2.6
Conclusion: the Physiological Significance of NSBE

From the above survey, it appears that the nitrate concentration of the SBE (NSBE) represents an integrated value of the NO_3^- exportation flux from the roots to the shoots, and an integrated value of the absorption flux during the days preceding the measurement.

10.3.3
Determination of a Critical Value of Nitrate Concentration in SBE

Figure 10.5 shows the relationship obtained in different field experiments between the NSBE and the index of nitrogen nutrition (NNI + 50), calculated for 50 degree-days (base 0) later (Justes 1993). We have indicated that the NSBE indicator responds earlier than NNI to a nitrogen deficiency: the delay in the latter has been estimated at 50 degree-days from results obtained by Gonzalez-Montaner (1987) and Justes (1993). Therefore, the NNI + 50 was estimated at day d plus 50 degree-days by linear interpolation, in relation to thermal time, between NNI measured at day d and NNI measured at the following sampling.

Justes (1993) obtained a significant correlation between NSBE and NNI + 50 for the different situations (various plant population densities, developmental stages, cultivars, soil types, countries, and growth conditions), ($r^2 = 0.754$; 175 degree of freedom). The covariance analysis (SAS Institute Inc 1987) indicated that

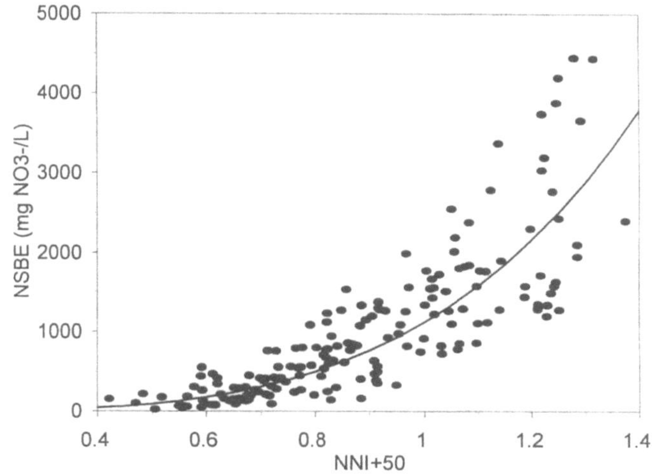

Fig. 10.5. Relationship between NSBE (measured at day d) and NNI + 50 which is NNI at day d plus 50 degree-days (+50°C days) for winter wheat crops (Justes 1993). Each *symbol* represents one point of measurement (mean of three replicates) and *curve* shows the global model (see text)

two factors have significant effects on this relationship: the plant density and growth stage. If we separate these factors into classes (two classes for plant density: normal and low, four classes for Feekes stage: 3 to 5, 6 to 7, 8 to 9 and 10 to 10.5), the complete model accounts for a greater proportion part of the variance (r^2 = 0.862; 176 degrees of freedom). The NSBE is significantly lower for the same NNI level when the plant density and the growth stage increase (Justes 1993).

Therefore, the NSBE threshold below which N is limiting growth (and the N fertilizer decision-making threshold) depends on the plant density and the developmental stage. In order to use the NSBE as an indicator of fertilizer management, we need to know the link between the wheat yield (or the number of grains m^{-2}) and the NNI.

We have shown previously (Chap. 2, this Vol. Cereal crops) that the grain number was not significantly different from the maximum grain number if NNI at anthesis was equal to or greater than 0.9. The threshold for N fertilizer decision-making has been defined as the nitrate concentration of SBE which existed for around days at 100°C (5 to 10 days in April and May in northern France) before the N deficiency could be detrimental, to allow for the time necessary for the N fertilizer action.

Complementary experiments allowed different threshold values to be established according to the pedoclimatic situation (three classes: north and middle of France, south and west of France and chalky soils of Champagne crayeuse) and cultivar (four classes according to earing precocity) (Justes et al. 1994a; Laurent et al. 1996).

For example, the NSBE threshold for fertilizer decision-making is:

- 1500 mg $NO_3^{-} l^{-1}$ for a standard density with the cultivar Soissons in the north of France at Feekes 6 stage, and 1000 mg $NO_3^{-} l^{-1}$ at Feekes 8–9 stage.

- 2300 mg $NO_3^- l^{-1}$ for a standard density with the cultivar Soissons in the chalky soils of the Champagne crayeuse at Feekes 6 stage, and 1800 mg $NO_3^- l^{-1}$ at Feekes 8–9 stage.
- 2000 mg $NO_3^- l^{-1}$ for a low density with the cultivar Soissons in the west of France at Feekes 6 stage, and 1500 mg $NO_3^- l^{-1}$ at Feekes 8/9 stage.

The method (balance-sheet + NSBE), called JUBIL, has now been commercialized in France by the company Challenge Agriculture. Farmers can use an interpretation grid, corresponding to crop density class and pedoclimatic situation, to interpret their value immediately. Consequently, a decision is proposed to omit or apply N fertilizer between the first node stage (Feekes 6) and booting stage (Feekes 8/9). The small case provided for the JUBIL method contains the equipment needed to make the nitrate test (NSBE indicator) under farm conditions: handle press, cutter, rapid test strips, reflectometer, interpretation grids, and the guidelines for converting the nitrate analysis into fertilizer recommendations.

10.3.4 Qualities and Limits of Utilization of the NSBE Indicator

The nitrate test is a good indicator for decision making because of its qualities (see Meynard et al., Chap. 9, this Vol.), particularly when an N fertilizer application is needed.

Due to the close relationship between NSBE and NNI + 50, and the lesser importance of root contribution to nitrate reduction, NSBE is closely related to the plant N status and thus its specificity is quite satisfactory. It is very sensitive in detecting the appearance of N deficiency. It may be used in all situations where soil inorganic N is primarily in nitrate form (which is the case in temperate climates).

The use of the NBSE indicator is easy for agronomists and also for farmers. The diagnosis of N nutrition is not influenced by the time of day that plants are sampled, or by large variations in crop growth rate, temperature, or solar radiation. Nevertheless, to use the NSBE correctly and give a robust diagnosis, it is necessary to take into account the plant density, the growth stage, the soil type, and the cultivar. Under these conditions, the nitrate test may be successfully applied across the large diversity of situations in France.

10.4 Validation of the JUBIL Method for Winter Wheat Crops

The JUBIL method was tested in France in 40 farmers' fields during the years 1991 to 1993. Various soil types, climatic conditions, winter wheat cultivars, and preceding crops were represented (Justes et al. 1994a).

In each case, four fertilizer N rates were applied (in two applications, at tillering and the beginning of stem elongation): X (balance-sheet rate), X-40

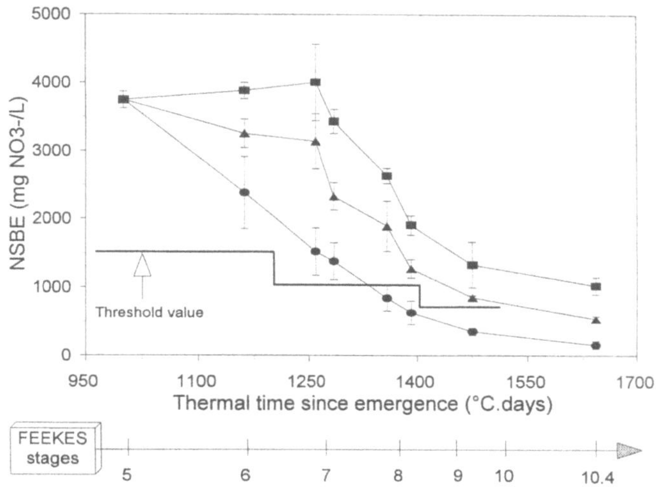

Fig. 10.6. NSBE versus thermal time from emergence for three N treatments. ■ X treatment; ● X-40 treatment; ▲ X-80 treatment (see text); *vertical bars* represent standard error (experiment at Eure et Loir, France in spring 1991; Justes 1993)

(JUBIL rate), X + 40 (safety rate) and 0 (no fertilizer). Sometimes, a rate of X-80 was also tested.

During stem elongation, the NSBE was measured in treatment X-40 (and sometimes X-80), and a third N dressing was applied on part of the experimental plot if NSBE was lower than the threshold value previously established, in order to maintain the yield objective.

Wheat yield components were analyzed in each treatment, including those which received an N dressing during stem elongation.

The mineral N remaining in soil at harvest was measured in six experiments in 1991 to evaluate the risk of nitrate leaching during the following winter (Justes 1993).

Figure 10.6 shows an example of the time course of NSBE on one site (Eure et Loir Veillard, 1991). The nitrate concentration decreased regularly during stem elongation and was clearly greater in the crops receiving more N. The NSBE measurements indicated that treatment X-80 became N deficient (lower than the threshold value) after the second node stage, and that N fertilizer was required to maintain the yield objective. On the other hand, the treatments X-40 and X did not require additional N, since NBSE always remained higher than the threshold value.

10.4.1 Evaluation of the JUBIL Method on Yield and Protein Content

We found that the NSBE in the X-40 treatment remained greater than the threshold value in 20 out of the 40 farmers' fields. In these fields, the JUBIL method recom-

Table 10.1. Average grain yields and protein contents observed in farm wheat fields (1991–1993)

NO_3^- concentration	> Threshold		< Threshold	
No. of sites	20		20	
N fertilizer rate	X-40	X	X-40 + 40	X
Yield ($t\,ha^{-1}$)	8.75a	8.84a	9.18a	9.05a
Protein content (%)	11.94b	12.58a	11.93a	11.67b

Letters a and b represent Newman-Keuls groups.

mended that extra N should not be applied and allowed a saving of $40\,kg\,N\,ha^{-1}$ compared to the balance-sheet rate X. In these X-40 treatments, the grain yield was not significantly different than in the conventional X treatments. The grain protein content was slightly reduced, on average by 0.6%, but stayed satisfactory. Thus, rate X-40 was the optimum amount of N fertilizer in terms of yield (Table 10.1).

In the 20 other fields, NSBE became lower than the threshold value during the measurement period, and a third dressing of $40\,kg\,N\,ha^{-1}$ was therefore applied. The total amount applied was the same in the X conventional treatment but three applications were made instead of two. Splitting N had no negative effect on yield; it even increased yield significantly in many cases, and slightly increased the protein content of grains.

10.4.2 Value of the JUBIL Method for Reducing the Risks of Nitrate Pollution

The amount of soil nitrate N at harvest was significantly reduced with the JUBIL method compared to the nominal X rate, in plots where the last dressing was not applied (X-40 rate) and even when the third $40\,kg\,N\,ha^{-1}$ dressing was applied (X-40[+40] rate) (Table 10.2).

The attraction of the JUBIL method to limit soil NO_3^- at harvest is more evident when one considers the "safety" rates (X + 40) frequently used by farmers. The nitrate in soil at harvest time is the form of N most susceptible to being leached during the next winter. By considering a mean of drainage of 150 mm (about the

Table 10.2. Nitrate-N in soil (depth: 0–90 cm) at harvest and nitrate concentration of water drainage calculated by the model of Burns (1976) for an average 150 mm drainage (mean in northern France)

Method	JUBIL	Balance sheet	Safety
N fertilizer rate	X-40 or X-40 + 40	X	X + 40
$N\text{-}NO_3^-$ ($kg\,ha^{-1}$)	26	36	50
NO_3^- in water ($mg\,l^{-1}$)	38	53	74

average of the last 20 years in the north of France) and using the model of Burns (1976), we can calculate that the mean concentration of nitrate in drained water would be 38, 53, and 74 mg $NO_3^- l^{-1}$ in treatments JUBIL, X, and X + 40, respectively. This shows the value of minimizing the residual nitrate in soil at harvest with this method.

These results have been confirmed by Laurent et al. (1996) for tests carried out in 1994 and 1995.

10.5
Use of Nitrate Concentration Analyses for Maize Crops

Following the work of Routchenko (1967) and Geyer and Marschner (1990), Plénet et al. (1994) developed a diagnostic tool for maize crops based on nitrate concentration in stem base using the concepts previously demonstrated in wheat (JUBIL method).

10.5.1
Measurement Protocol of Nitrate Test for Maize

The operative protocol of the NSBE in maize is as follows (Plénet et al. 1994):

- Ten plants on four to six plots are sampled in early morning. The plants are cut just above the soil (first visible internode) and at the third internode level of stem, so as to keep only the second internode without the leaf sheaths. The samples are preserved in an ice box during transport to the laboratory or the farm. Before the 12-13-leaf stage, the plant tissues selected correspond to "*pseudo*-stem".
- Disks of 1–2 cm thickness are sampled at the middle of the second stem internode and the extraction is carried out with a manual press.
- The nitrate concentration is measured using a colorimetric method with an autoanalyzer or with rapid test strips and a reflectometer.

10.5.2
Time Course of the NSBE in the Maize Crops: Elements of Physiological Significance

An example of the time course of SBE nitrate concentration during maize crop growth is shown in Fig. 10.7, and this can be compared to NNI shown in Fig. 10.8. In these experiments fertilizer N is applied twice, with 30 kg N ha^{-1} at planting and the remainder at the 8-9-leaf stage. The second application produces a large increase in the NSBE, except with the 30N treatment. This can be interpreted as the result of root nitrate reduction capacity being exceeded when the nitrate supply is very large (Gojon et al. 1986). In this case, the majority of nitrate absorbed by the roots is translocated to the shoots, where the reduction capacity and supply of carbohydrates are not limiting under favorable conditions of solar radiation.

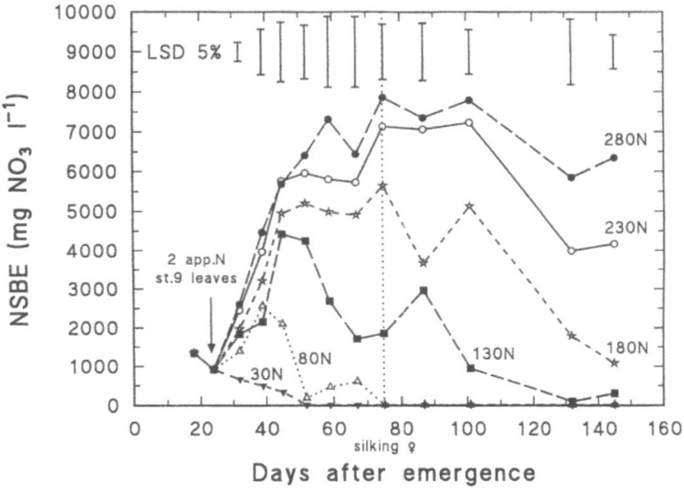

Fig. 10.7. Effects on N application rates (six N rates from 30 to 280 kg N ha^{-1}) on the time course of NSBE (mg $NO_3^{-1}l^{-1}$) during the growing season at Onard 1992 (France). Data are the mean of four replicates

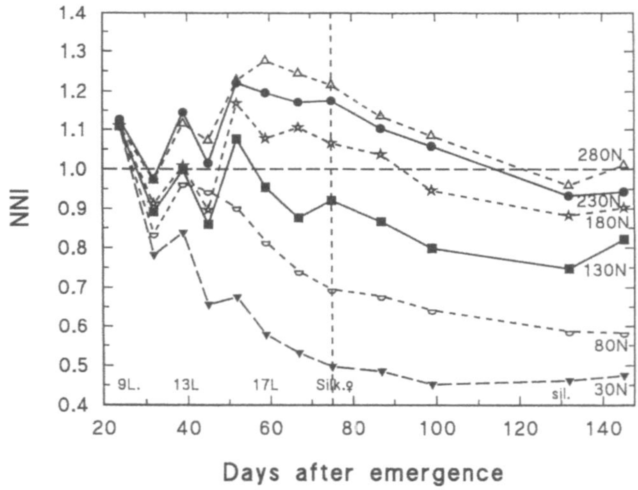

Fig. 10.8. Time course of NNI (NNI = $\%N_{act}/\%N_c$ with $\%N_c = 3.40\ W^{-0.37}$) as a function of N rates (six N rates from 30 to 280 kg N ha^{-1}) at Onard 1992 (France). Data are the mean of four replicates

When the N supply is limiting, which produces a decrease in NNI (Fig. 10.8), the NSBE declines quickly and approaches zero. Even if the absence of nitrate in the basal stem does not signify the cessation of nitrogen absorption by the roots, it does indicate that the flux of reduced N is inadequate for N growth demand, at least for the 10-leaf stage and flowering. During this period, NSBE is a very precise indicator for detecting the appearance of N deficiency, showing a response 6 days

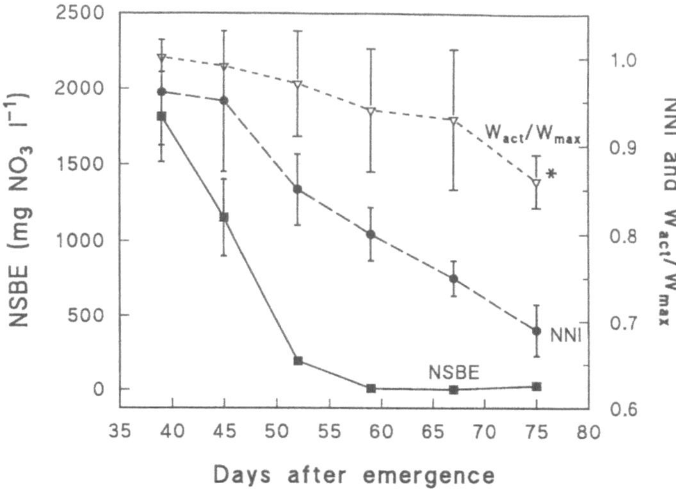

Fig. 10.9. Comparison of the time course of NSBE ($mg\,NO_3^{-1}\,l^{-1}$), the NNI ($NNI = \%N_{act}/\%N_c$ with $\%N_c = 3.40\,W^{-0.37}$) and the relative shoot dry matter (W_{act}/W_{max}) on the treatment 80N ($80\,kg\,N\,ha^{-1}$) during the period from 39 days after emergence (13-leaf stage) to 75 days after emergence (silking stage). *Error bars* show standard error of the mean of three replicates and * shoot dry matter is significantly different at $P < 0.05$

(65°C-days) before NNI decreased, and 28 days (350°C-days) before a significant reduction in shoot growth (Fig. 10.9).

After flowering plus 20 days, the stem base is primarily a storage organ for excess nitrate. This storage can reach up to $35\,kg\,N\,ha^{-1}$ ($8\,t\,DM\,ha^{-1}$ of stems at 20% DM with a gradient of nitrate concentration between 0 to $10\,000\,mg\,NO_3^-\,l^{-1}$ from the base of the stem to its top), which represents a considerable amount compared to the 7 to $8\,kg\,N\,ha^{-1}$ which are necessary to make $1\,t\,DM\,ha^{-1}$ during this period. However, this storage pool does not always seem to be used well during the grain-filling period. This difficulty of the plants in using stem nitrate at the end of the grain accumulation period was reported by Below et al. (1981). It seems to be related to the absence of nitrate reductase activity in the stem at 30 days after anthesis (Ta 1991). On the other hand, high nitrate concentrations in SBE at the silage and maturity stages give a good indication of excessive N supply, as has been shown by Binford et al. (1992) on extracts carried out on dried stems.

Use of nitrate test on maize seems possible from the 10-leaf stage to the 20 days postflowering stage, at the latest. However, in areas where N is applied twice, it is necessary to carry out several measurements of the nitrate concentration in the young plants, so as to have a precise diagnosis of plant nutrition before the 14-leaf stage. The agronomic value of the nitrate test at the juvenile stage (influence of absorption of ammonium in cold soils) and after flowering (decrease of nitrate reductase activities and increase of proteolytic activities) must still be demonstrated before the test is used in these cases.

10.5.3
Relation Between the Nitrate Test and Nitrogen Nutrition Index: Determination of Nitrate Concentration Threshold for Maize Crops

The nitrate test is a partial indicator of plant N status, as it is carried out on a small part of the plant and takes into account only nitrates, which represent a small part of the total N of the plant. Therefore we have compared the nitrate test with the NNI in maize crops, which is an indicator of the overall N status of the crop. Figure 10.10 shows the NSBE in relation to the NNI at different stages of maize crop on several locations in France from 1992 to 1994 (Plénet 1995, and unpubl. data).

The NSBE is low when NNI is lower than 1. This shows that nitrate reduction must occur mainly in the roots when the N supply is less than the demand. The N is transported from root to shoot in the form of amino acids and there is no nitrate accumulation in the stem base. On the other hand, the NSBE tends to increase strongly when NNI > 1; but there is no linear relation between the two indicators: the nitrate concentration may range from 1000 to 8000 mg l^{-1} for values of NNI between 1 and 1.6. In this case, nitrate storage in the stem base indicates that the supply is greater than the demand. The amount of nitrate in SBE depends on the availability of carbohydrates and of reduced energy carriers produced by the light

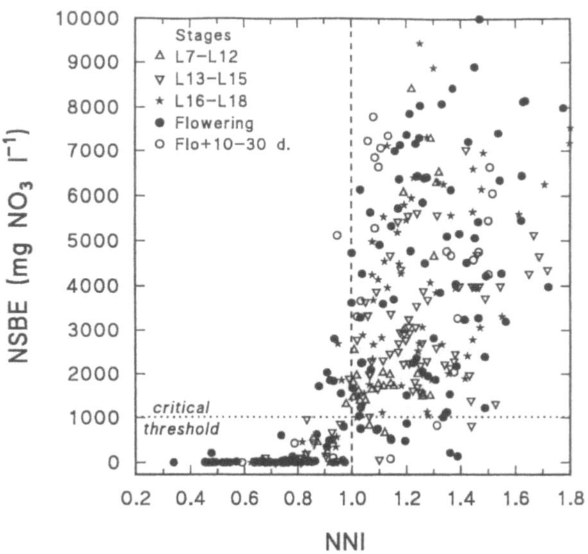

Fig. 10.10. Relationship between NSBE (mg NO_3^{-1} l^{-1}) and NNI (NNI = $\%N_{act}/\%N_c$ with $\%N_c$ = 3.40 $W^{-0.37}$). The two indicators of N status are measured at the same date during the period from 7-leaf stage to 20 days after flowering. Data of 20 experiments realized in France from 1992 to 1994. The critical threshold of 1000 mg NO_3^{-1} l^{-1} was determined in using the procedure presented in the text and Table 10.3 ($n = 352$)

reactions, which relate to environmental conditions (Hageman 1986; Touraine et al. 1994).

To determine a nitrate concentration threshold, we have used NNI = 1, because it corresponds to the minimal level of N nutrition that allows the maximal growth at a given date. Furthermore we have previously shown (Chap. 5, this Vol.) that the number of grains per unit area will not be significantly different from the maximum when NNI is equal to or greater than 1 at flowering (Plénet 1995). The relationship between NNI and the critical nitrate concentration ($NO_3^-{}_c$) in SBE must be described by: NNI < 1, NO_3^- < $NO_3^-{}_c$ and NNI > 1, NO_3^- > $NO_3^-{}_c$. The experimental variability introduces two sorts of errors: error A when NNI < 1, but NO_3^- > $NO_3^-{}_c$ and error B when NNI > 1, but NO_3^- < $NO_3^-{}_c$. For every stage, we have calculated the probability of realization of the two types of errors for different nitrate concentrations (Table 10.3). The nitrate concentration threshold which maintain error A at less than 5% probability is between 750 and 1250 mg NO_3^- l^{-1}, according to the growth stage. The critical threshold of 1000 mg NO_3^{-1} l^{-1} has been selected because it minimizes the two errors. To confirm this threshold, the same analysis was made when using the relative shoot biomass (DM/DM_{max}) measured 14 days after the nitrate test (short-term prediction). The purpose of this analysis was to check whether dry matter accumulation was significantly limited by N, and when the NSBE was higher than 1000 mg NO_3^{-1} l^{-1}. Except for one case, when the nitrate test result was greater than the threshold value, the shoot biomass was not limited (data not presented).

To use the nitrate test as a tool in N fertilization management of maize crops, a short-term prognosis must be carried out integrating the time necessary for farmers to carry out the N fertilizer application and for the N fertilizer to be available for the plant (at least 7–15 days). The critical concentration that is applied corresponds to an intervention threshold. Based on 20 experiments, we adopted an intervention threshold of 1500 mg NO_3^{-1} l^{-1}, which corresponds to the

Table 10.3. Probability of two diagnostic error types as a function of different nitrate concentrations in stem base extract at different stages in maize (L: number of visible leaves)

Stage	Number data	Nitrate concentration in SBE (mg NO_3^- l^{-1})											
		500		750		1000		1250		1500		2000	
		A[a]	B	A	B	A	B	A	B	A	B	A	B
		(%)											
7L-12L	32	6.3	0.0	3.1	3.1	3.1	6.3	3.1	6.3	0.0	21.9	0.0	56.3
13L-15L	77	5.2	1.3	3.9	1.3	1.3	2.6	1.3	3.9	1.3	7.8	0.0	15.6
16L-18L	82	3.7	0.0	2.4	0.0	2.4	0.0	2.4	4.9	2.4	4.9	1.2	12.2
Flowering	122	8.2	2.5	5.7	4.1	4.9	6.6	4.9	10.7	4.9	10.7	1.6	14.8
Flowering + 10 to 30 days	39	2.6	2.6	2.6	5.1	2.6	7.7	2.6	10.3	2.6	10.3	2.6	12.8
Total	352	5.7	1.4	4.0	2.6	3.1	4.3	3.1	7.4	2.8	9.7	1.1	17.9

[a] A Error: NNI < 1, but NO_3^- > $NO_3^-{}_c$ and B Error: NNI > 1, but NO_3^- < $NO_3^-{}_c$.

critical threshold of 1000 mg NO_3^{-1} l^{-1}, to which we have added 500 mg NO_3^{-1} l^{-1} to allow for the time necessary for the N fertilizer action (about 7 days). We advise that the nitrate test be used during the period from the 12-leaf stage to flowering. The nitrate indicator permits only a short-term prognosis, as the threshold of intervention corresponds to the nitrate concentration above which the N nutrition will not penalize growth in the 15 days following the date of the nitrate test. This threshold of intervention is being tested for maize crops at the national level in France in 1996.

10.5.4
Use and Perspectives for Maize Crops

The nitrate test proves to be a rapid and reliable diagnostic tool for maize crops, which can be used to evaluate a strategy of fertilization, as for the JUBIL method proposed for winter wheat. The management of fertilizer application in maize is based on four steps:

1. Determination by the balance sheet method of the total N fertilizer requirements: rate X (kg N ha^{-1}).
2. Application of the amount equal to X minus 50 kg N ha^{-1} at the start of cropping (planting plus 8-leaf stage).
3. Monitoring of crop N status with the plant nitrate test from the 12-leaf stage to flowering stage (three to four measuring dates).
4. If the nitrate test is lower than 1500 mg NO_3^{-1} l^{-1} during one of the measurement dates, a supplementary dose of 50 kg N ha^{-1} is applied in the following week.

Nevertheless, in view of the height of the maize crop, such management of N fertilizer applications by monitoring crop N status necessarily implies the use of specific techniques to spread the supplementary fertilizer on the crop after the 12-leaf stage: N applied with irrigation water or a high-clearance vehicle equipped with soil injectors. The large areas of irrigated maize cropping high leaching potential and/or low capacity for water retention (systems which are vulnerable to nitrate pollution), are an incentive to perfect the nitrate test so that the N supply is well adjusted to the growth needs.

In those areas where it is not possible to carry out the supplementary N application, the validation of a rapid diagnostic tool proves to be essential for a posteriori analysis of the experimental results, so as to increase the precision of reference data for each pedoclimatic situation. The risks to the environment will thus be progressively reduced.

10.6
Conclusion

This chapter has shown the interest of linking a simple predictive model of N fertilization (balance-sheet method) and a plant indicator (NBSE) to optimize N fertilizer for a given target yield and to reduce nitrate pollution as much as possible. The NBSE has shown its value for detecting the appearance of N deficiency and for deciding on the need for an extra N dressing during stem elongation in wheat and maize. However, it cannot be used to verify crop N status after an N fertilizer application carried out during stem elongation of winter wheat.

The use of a plant indicator in the decision-making rules for N fertilization may allow farmers to change the management of risks. To avoid N deficiency, farmers usually apply more fertilizer N than the amount calculated by the balance-sheet method (i.e., overestimation of the target yield). Thus, farmers prefer to obtain a slight excess of nitrogen supply which is less detrimental to yield (e.g., lodging, plant diseases, etc.) than N deficiency. In these conditions, there is no a priori risk of N deficiency to ensure the variable margin that is aimed for. This attitude is potentially bad for the environment. Models to predict N fertilization (e.g., the balance sheet method) do not allow a complete change in this attitude. On the other hand, the nitrate test gives information to apply a last N dressing; it is possible then to apply a priori a reduced amount of fertilizer N without any consequence for yield. Hence, risks of overapplications of fertilizer are strongly reduced without any risk of yield decrease due to N deficiency.

The JUBIL method has been implemented for winter wheat in 1994 by extension services (ITCF) with farmers' groups, to test the method's feasibility on a large scale in France. It received a promising reception because farmers have easily adopted the technique and used it on a large scale (200 000 ha of winter wheat were managed in 1995 using the JUBIL method). Its use showed be made a priority on the fields where N fertilizer rate is more difficult to adjust (after fallow, organic application, legume crops, etc.). It could be, in a second phase, extended to a larger number of cereal fields.

Its adaptation to other crop species (rapeseed, potatoes, sugarbeet, etc.) should be studied, particularly for summer crops because the net mineralization component of the N balance is more important and variable according to climatic conditions than is the case for winter crops. However, practical problems such as those evoked for maize could be a brake to extending the use of the JUBIL method. Indeed, N dressings applied in summer are not easily available for root absorption: in this case irrigation is necessary.

Acknowledgments. We wish to thank François Laurent (ITCF) and Philippe Desvignes (AGPM), who authorized the use of experimental data obtained by ITCF (French Technical Institute for Cereal and Forage crops) and AGPM (French Syndical and Technical Institute for Maize crops).

References

Beevers L, Hageman HR (1980) Nitrate and nitrite reduction. In: Miflin BJ (ed) The biochemistry of plants, vol 5, Amino acids and derivatives. Academic Press, New York, pp 115–168

Below FE, Christensen LE, Reed AJ, Hageman RH (1981) Availability of reduced N and carbohydrates for ear development of maize. Plant Physiol 68:1186–1190

Beringer H, Hess G (1979) Feasibility of plant analysis for estimating late N requirements of winter wheat. Landwirtsch Forsch 32:384–394

Binford GD, Blackmer AM, El-Hout NM (1990) Tissue test for excess nitrogen during corn production. Agron J 82:124–129

Binford GD, Blackmer AM, Meese BG (1992) Optimal concentrations of nitrate in cornstalks at maturity. Agron J 84:881–887

Burns IG (1976) Equations to predict the leaching of nitrate uniformly incorporated to a known depth or uniformly distributed throughout a soil profile. J Agric Sci Camb 86:305–313

Coïc Y (1956) La nutrition et la fertilisation du blé d'hiver. Les besoins en azote du blé d'hiver. Conséquences agronomiques. Ann Agron 7:115–131

Cooper HD, Clarkson DT, Johnston MG, Whiteway JN, Loughman BC (1986a) Cycling of amino-nitrogen between shoots and roots in wheat seedlings. In: Lambers H, Neeteson JJ, Stulen I (eds) Fundamental ecological and agricultural aspects of nitrogen metabolism in higher plants. Proc Symp April 1985, Univ of Groninger and Inst for Soil Fertility, Haren, pp 97–100

Cooper HD, Clarkson DT, Ponting HE, Loughman BC (1986b) Nitrogen assimilation in field-grown winter wheat: Direct measurements of nitrate reduction in roots using ^{15}N. In: Lambers H, Neeteson JJ, Stulen I (eds) Fundamental ecological and agricultural aspects of nitrogen metabolism in higher plants. Proc Symp April 1985, Univ of Groninger and Inst for Soil Fertility, Haren, pp 395–398

Darby RJ, Widdowson FV, Bird E, Hewitt MV (1986) The relationship of soil mineral NO_3-N with stem NO_3-N concentration, and of fertilizer-N with the amount of nitrogen taken up by winter wheat, in experiments testing nitrogen fertilizer in combination with aphicide and fungicides, from 1980 to 1982. J Agric Sci Camb 106:497–507

Fox RH, Roth GW, Iversen KV, Piekielek WP (1989) Soil and tissue nitrate tests compared for predicting soil nitrogen availability to corn. Agron J 81:971–974

Gardner BR, Jackson EB (1976) Fertilization, nutrient composition, and yield relationships in irrigated spring wheat. Agron J 68:75–78

Geyer B, Marschner H (1990) Charakterisierung des Stickstoffversorgungsgrades bei Mais mit Hilfe des Nitra-schnelltests. Z Pflanzenernähr Bodenkd 153:341–348

Gojon A, Soussana JF, Passama L, Robin P (1986) Nitrate reduction in roots and shoots of barley and corn seedlings: I. 15N study. Plant Physiol 82:254–260

Gonzalez-Montaner JH (1987) Deux outils d'analyse de la réponse à la fertilisation azotée: les composantes du rendement et la teneur en nitrates des organes végétaux. Thèse de Docteur-Ingénieur INA P-G, Paris

Hageman RH (1986) Nitrate metabolism in roots and leaves. In: Shannon JC, Knievel DP, Boyer CD (eds) Regulation of carbon and nitrogen reduction and utilization in maize. Proc 1st annual Penn State Symp in Plant Physiology. Pennsylvania State University, American Society of Plant Physiologists, Bethesda, 372 pp

Hébert J (1969) La fumure azotée du blé tendre d'hiver. Bull Tech Inf Minist Agric Fr (BTI) 224:755–766

Iversen KV, Fox RH, Piekielek WP (1985) Diurnal, shade and hybrid effects on nitrate content of young corn stalks. Commun Soil Sci Plant Anal 16:837–852

Justes E (1992) Diagnosis of nitrogen nutrition status of winter wheat crop by "nitrate test": relationship with the nitrogen dilution curve during the growth. In: Scaife A (ed) Proc 2nd ESA Congr, Warwick Univ, Waiwick, pp 262–263

Justes E (1993) Diagnostic de la nutrition azotée du blé à partir de la teneur en nitrate de la base de la tige. Application au raisonnement de la fertilisation. Thèse de Docteur INA P-G, Paris, 227 pp + annexes

Justes E, Meynard J-M, Mary B, Laurent F (1994a) JUBIL: a new method of conducting the nitrogen fertilization in winter wheat crops. Proc 3rd ESA Congr, Padova, pp 490–491

Justes E, Mary B, Meynard J-M, Machet J-M, Thelier-Huché L (1994b) Determination of a critical nitrogen dilution curve for winter wheat crops. Ann Bot 74:397–407

Knowles TC, Doerge TA, Ottman MJ (1991) Improved nitrogen management in irrigated durum wheat using stem nitrate analysis: II. Interpretation of nitrate-nitrogen concentrations. Agron J 83:353–356

Large EC (1954) Growth stages in cereals. Illustration of the Feekes scale. Plant Pathol 3:129

Laurent F, Justes E, Gate P (1996) JUBIL 1996: La méthode s'affine. Perspect Agric 214:63–74

MacClenahan EJ, Killorn R (1988) Relationship between basal corn stem nitrate content at V6 growth stage and grain yield. J Prod Agric 1:322–326

Machet JM, Dubrulle P, Louis P (1990) Azobil: a computer program for fertilizer N recommendations based on a predictive balance sheet method. Proc 1st ESA Congr, Paris, Session 2, p 21

Meynard JM, Boiffin J, Caneill J, Sebillotte M (1981) Elaboration et fertilisation azotée du blé d'hiver en Champagne crayeuse. II Types de réponse à la fumure azotée et application de la méthode du bilan prévisionnel. Agronomie 1:795–806

Papastylianou I, Graham RD, Puckridge DW (1982) The diagnosis of nitrogen deficiency in wheat by means of a critical nitrate concentration in stem bases. Commun Soil Sci Plant Anal 13:473–485

Papastylianou I, Graham RD, Puckridge DW (1984) Diagnosis of the nitrogen status of wheat at tillering and prognosis for maximal grain yield. Commun Soil Sci Plant Anal 15:1423–1436

Plénet D (1995) Fonctionnement des cultures de maïs sous contrainte azotée. Détermination et application d'un indice de nutrition. Thèse Docteur Institut National Polytechnique de Lorraine, Université de Nancy, Nancy, 247 pp

Plénet D, Desvignes Ph (1993) Diagnostic de la nutrition azotée d'une culture de maïs. Mise au point d'un test basé sur la teneur en nitrate des jus extraits de la base des tiges du maïs. Essais de Onard 1992 et 1993. Experimental report INRA-AGPM, INRA Bordeaux, 29 pp plus annexes

Plénet D, Desvignes Ph, Kojfer L (1994) Evolution de la concentration en nitrate dans la base des tiges de maïs comme indicateur précoce d'une déficience en azote. CR Acad Agri Fr 80(2):129–138

Rémy JC (1981) Etat actuel et pérspectives de la mise en oeuvre des techniques de prévision de la fumure azotée. CR Acad Agri Fr 67:859–872

Rémy JC, Hébert J (1977) Le devenir des engrais azotés dans le sol. CR Acad Agric Fr 63:700–710

Routchenko W (1967) Appréciation des conditions de la nutrition minérale des plantes basée sur l'analyse des sucs extraits des tissus conducteurs. Ann Agron 18:361–402

SAS Institute Inc (1987) SAS/STAT guide for personal computers, Version 6 Edition. SAS Institute Inc, Cary, North Carolina

Scaife A, Stevens KL (1983) Monitoring sap nitrate in vegetable crops: comparison of test strips with electrode methods, and effects of time of day and leaf position. Commun Soil Sci Plant Anal 14:761–771

Ta CT (1991) Nitrogen metabolism in the stalk tissue of maize. Plant Physiol 97:1375–1380

Touraine B, Clarkson DT, Muller B (1994) Regulation of nitrate uptake at the whole plant level. P11–30. In: Roy J, Garnier E (eds) A whole plant perspective on carbon-nitrogen interactions. SPB Academic Publishing, The Hague

Wehrmann J, Scharpf HC, Böhmer M, Wollring J (1982) Determination of nitrogen fertilizer requirements by nitrate analysis of the soil and of the plant. Proc 9th Int Plant Nutrition Coll 091, vol 2. Warwick Univ, Warwick pp 702–709

CHAPTER 11

Leaf N Content as an Indicator of Crop N Nutrition Status

G. Lemaire[1], D. Plénet[2], and D. Grindlay[1]

Contents

11.1 Introduction

The major difficulty in the direct use of the nitrogen nutrition index (NNI) as an indicator of crop N status at the farm level is the need to determine the actual crop mass to calculate the NNI value. As shown in Chapter 2, the NNI can be a valuable reference tool for experimental conditions where the determination of crop mass can be easily made, but it cannot be considered as an operational tool for crop diagnosis. It therefore seems necessary to determine effective and simple procedures for the diagnosis of crop N status which do not require a determination of crop mass, but which are closely related to the NNI of the crop.

11.2 Can Leaf N Content Be Used as an Indicator of Crop N Status?

Many attempts have been made to relate the concentration of N in leaf tissue dry matter to plant N status as a diagnostic tool (Lagatu and Maume 1926; Ulrich 1943; Smith 1962). Such an approach makes physiological sense, since much of the demand of a plant for N is determined by the photosynthetic function of the leaves.

[1] INRA Station d'Ecophysiologie des Plantes Fourragères, 86600 Lusignan, France
[2] INRA Station d'Agronomie, Domaine de la Grande Ferrade, BP 81, 33883 Villenave d'Ornon Cédex, France

Diagnosis of the Nitrogen Status in Crops
G. Lemaire (ed)

Moreover, leaf N content is correlated with leaf extension rate (Gastal et al. 1992). Thus, leaf N content determines not only the amount of CO_2 assimilation per unit leaf area, but also relates closely to the rate of use of C for leaf expansion, and hence the amount of light intercepted by the plant.

A system of foliar diagnosis based on leaf N concentration and extended to other elements such as P, K, and Mg, was developed by Beaufils (1973) and Walworth and Summer (1987) as part of the DRIS method (diagnosis and recommended integrated system). This system has been widely used for sugar cane (El Wali and Gascho 1984), and for some other crops such as soybean (Bever 1986), maize (Edwall et al. 1985), and potatoes (Meldal-Johnsen and Summer 1980). Despite the large number of studies concerning foliar diagnosis in the literature, the technique has not been widely adopted by agronomists as a routine procedure. The reason for this remains the large divergence between authors in the determination of "critical concentration", due to the problem of sampling of leaves in relation to their age and position inside the canopy.

As light interception is an area-based phenomenon, it may be preferable to express the N content of leaves on a leaf area basis. Sinclair and Horie (1989) reviewed a number of studies that showed a high correlation between the light-saturated CO_2 assimilation rate of leaves and their content of N per unit leaf area. Muchow and Sinclair (1994) found a hyperbolic relationship between leaf photosynthetic capacity and leaf N per unit area in maize and sorghum. From this relationship it can be estimated that in these C_4 crops a leaf N content of around $1.5\,g\,m^{-2}$, gives the maximum potential leaf photosynthesis under light-saturating conditions. Below this value, the photosynthetic capacity of the leaves decreases as their N content decreases, and becomes virtually zero for a leaf N content of around $0.2–0.3\,g\,m^{-2}$. Similar relationships between leaf N content and leaf photosynthesis at saturating irradiance have been found for several other species depending on their metabolic pathways, C_3 or C_4 (Evans 1989; Anten at al. 1995). According to the theory developed in Chapter 1, leaf N content per unit leaf area represents the N concentration of the metabolic part of the plant, and therefore governs the plant's metabolic activity.

As noted in Chapter 1, in dense crops the N content of leaves (and hence their photosynthetic capacity) decreases in parallel with light extinction down the canopy (Charles-Edwards et al. 1987; Werger and Hirose 1991; Anten et al. 1995). For lucerne, Lemaire et al. (1991) showed that leaf N content per unit leaf area, N_L, declined exponentially with depth z down the canopy:

$$N_L = a\,e^{-bz}. \qquad (1)$$

The relative stability of the two coefficients a and b of Eq. (1) at different stages of regrowth of a lucerne stand is demonstrated in Table 11.1. This shows that during the regrowth, new leaves appeared at the top of the canopy, in the well-illuminated layer, at a relatively constant N_L of $2.45–2.82\,g\,m^{-2}$, while the N_L of older leaves declined as they became progressively more shaded. The maintenance of a relative constant N_L at the top of the canopy occurred despite the general decrease in plant

Table 11.1. Values of the coefficients of the relationship $N_z = ae^{bz}$ between nitrogen content per leaf area (N_z) and the depth (z) of leaf within a lucerne canopy

Date	a $gN m^{-2}$	b
29 April	2.45	0.021
19 May	2.64	0.021
22 June	2.73	0.024
30 June	2.82	0.022

a represents leaf N content ($gN m^{-2}$) at the top of the canopy, b is an adjusted parameter.

N% with increasing crop mass, as described by Eq. (1) in Chapter 1, $N\% = a(W)^{-b}$. It should be possible to determine a critical N_L, i.e., the minimum leaf N per unit leaf area at the top of the canopy necessary to maintain maximum crop growth rate. Unlike critical plant N% which decreases with increasing crop mass, critical N_L could be considered as independent of crop biomass changes during a given period of growth.

If N_L at a given point in the canopy changes with N supply, and there is evidence that this is so (Sinclair and Horie 1989; Muchow and Sinclair 1994), then the ratio between actual N_L and critical N_L could be substituted for NNI as an indicator of crop N status, without the need to measure crop biomass. The data on lucerne reported by Lemaire et al. (1991) can provide only an indication of the independency of N_L in relation to crop mass. The absence of a large variation in crop N status below the optimum in lucerne does not allow a precise determination of critical N_L.

11.3
Variation in N_L with Leaf Position and N Supply in a Maize Crop

An experiment was carried out in 1994 with maize crops receiving a range of N fertilizer rates at two sites, Onard and Lusignan, in the southwest of France. The aim was to investigate in more depth the critical N_L and its independency of crop biomass at various stages of growth. At each sampling date, the NNI was calculated according to the procedure described in Chapter 5 by determination of actual plant N% and crop biomass from sequential plant samples. At the same time, N_L of selected leaf numbers was determined for five plants per plot, taking the mean of three or six punched disks (each $1.96\,cm^2$ in area) per lamina, taking care to exclude the midrib. All analyses were for total N content.

Figure 11.1 represents the time course of NNI for the six levels of N supply in one of the two experiments, at Onard. The range of NNI values at the time of female flowering in this experiment was 0.6–1.3, which indicates the large contrast in crop N status that was obtained.

11.3.1
Time Course of N_L Profiles Within the Canopy

The time course of N_L ($g\,m^{-2}$) in selected leaf numbers for the different N treatments at Onard is shown in Fig. 11.2. It can be seen that for each leaf a maximum N_L was reached when full expansion had occurred, and N_L then declined gradually with leaf age and/or progressive shading by newer leaves above. For treatments corresponding to N deficiency, as indicated by NNI values less than 1 in Fig. 11.1, there was a reduction in N_L on a given date. However, for treatments with NNI values greater than 1, no further increase in N_L was observed, indicating that there was little or no luxury accumulation of N in the leaves.

Figure 11.3 shows profiles of N_L down the canopy on different dates at Onard. The range in N_L values down the canopy in plants well supplied with N, from 1.6 down to $0.6\,g\,m^{-2}$, was very similar to that obtained by Anten et al. (1995) in sorghum (although these authors sampled whole leaves, not just lamina tissue). It can be seen in Fig. 11.3 that the N_L profiles for the different N treatments were relatively parallel, with an effect of N supply on N_L at all positions in the canopy. For each leaf number, N_L discriminated well between treatments with NNI > 1 and treatments with NNI < 1. Figure 11.4 shows the similar results obtained in the experiment at Lusignan, where two cultivars were used. In both experiments it was the last visible collar leaf which showed the maximum N_L at each sampling date. Leaves which were still expanding had not yet reached their maximum N content, perhaps because they were erect and received a lower irradiance than did mature

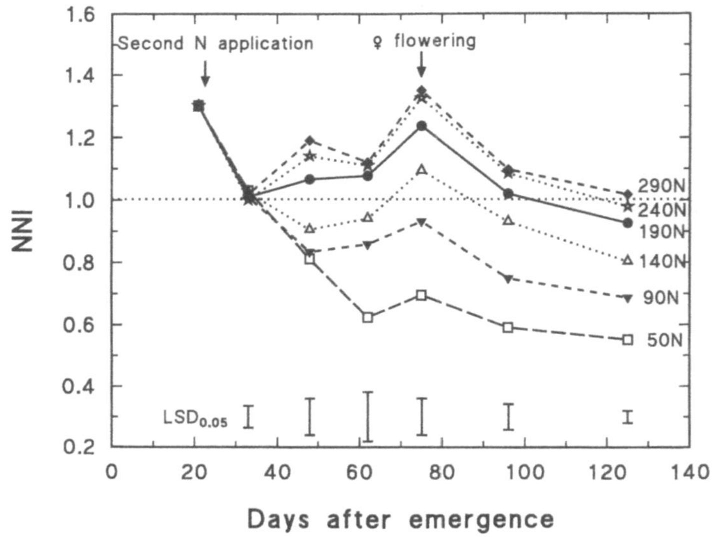

Fig. 11.1. Time course of nitrogen nutrition index (*NNI*) in maize, cultivar Volga, grown at Onard in 1994 with six N fertilizer application rates (indicated in $kg\,N\,ha^{-1}$ on the figure); $50\,kg\,N\,ha^{-1}$ was applied at sowing for all treatments

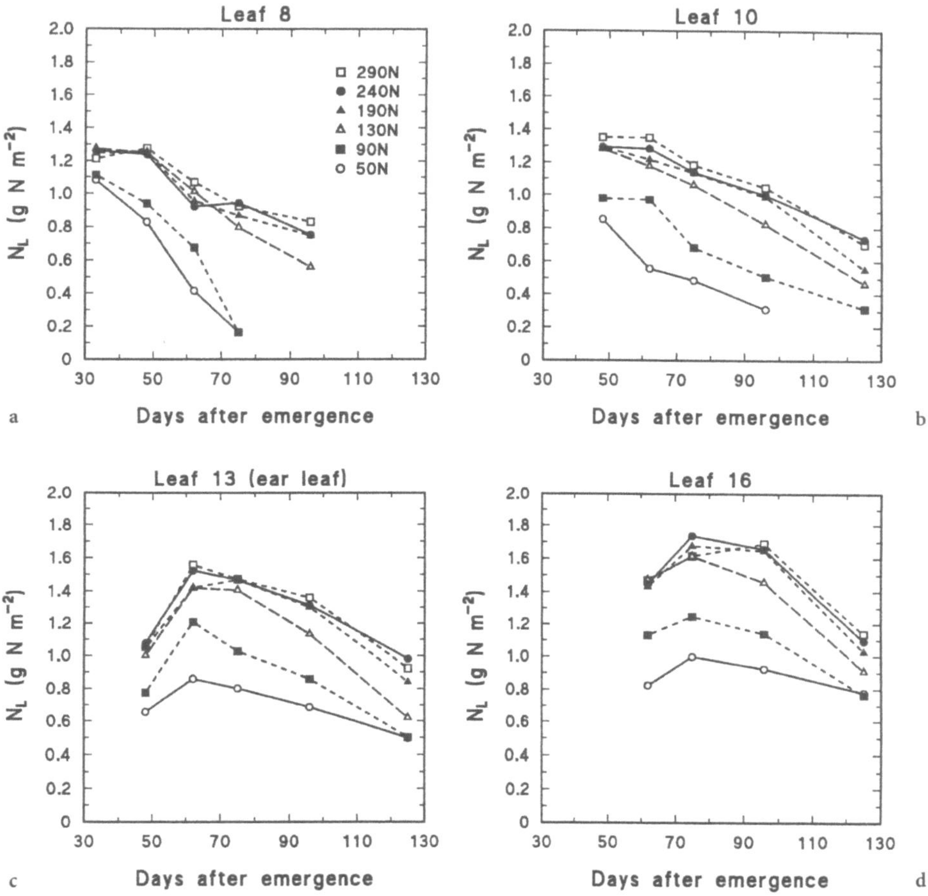

Fig. 11.2a–d. Time-course of N per unit area (N_L) in selected leaves of maize, cultivar Volga, grown at Onard in 1994 with six N fertilizer application rates (indicated in kg N ha^{-1} in **a**). Leaves are denoted as leaf number counting from the shoot base

leaves. After maturity, the N content of leaves declined progressively as they aged and became more shaded. Thus the N_L of the last visible collar leaf seems suitable to represent plant N status in the period before flowering.

11.3.2
Maximum N_L in the Canopy in Relation to Crop N Status

Figure 5a and b represents the time course of N_L measured in the last visible collar leaf in the two experiments, that is in mature, unshaded leaves at the canopy top. The maximum N_L value obtained for treatments with NNI > 1 showed some

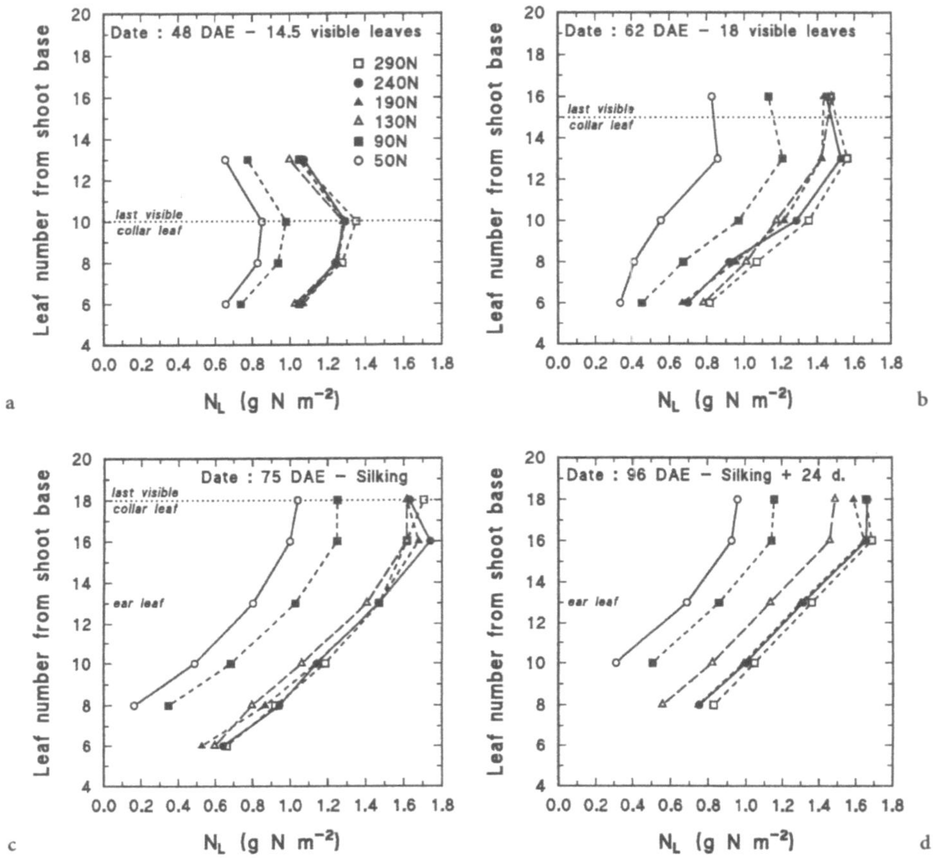

Fig. 11.3a–d. Profiles of N per unit area (N_L) down the canopies of maize crops, cultivar Volga, grown at Onard in 1994 with six N fertilizer application rates (indicated in $kg N ha^{-1}$ in **a**). Data for four sampling dates, expressed as days after emergence (DAE)

variation around the optimum value of $1.5 g m^{-2}$ proposed by Sinclair and Horie (1989) for maize and sorghum; but the variations in maximum N_L with time seemed similar for the two experiments, with a decline from 1.5 to $1.3 g m^{-2}$ during the first 50 days followed by an increase from 1.3 to $1.6 g m^{-2}$ until silking. After silking, the decrease in N_L was presumably due to leaf aging and remobilization of N to the grain.

The maximum N_L at each data could not be determined precisely because only every second leaf number on the stem was sampled. Hence, we did not necessarily include the last visible collar leaf on each date, and the maximum N_L may sometimes have been underestimated, which could be the cause of some of the variation seen in Fig. 11.5. The data are therefore not sufficient to conclude on the constancy of the maximum N_L, our critical N_L, and further experiments will be needed where the last visible collar leaf is always included in the sampling procedure.

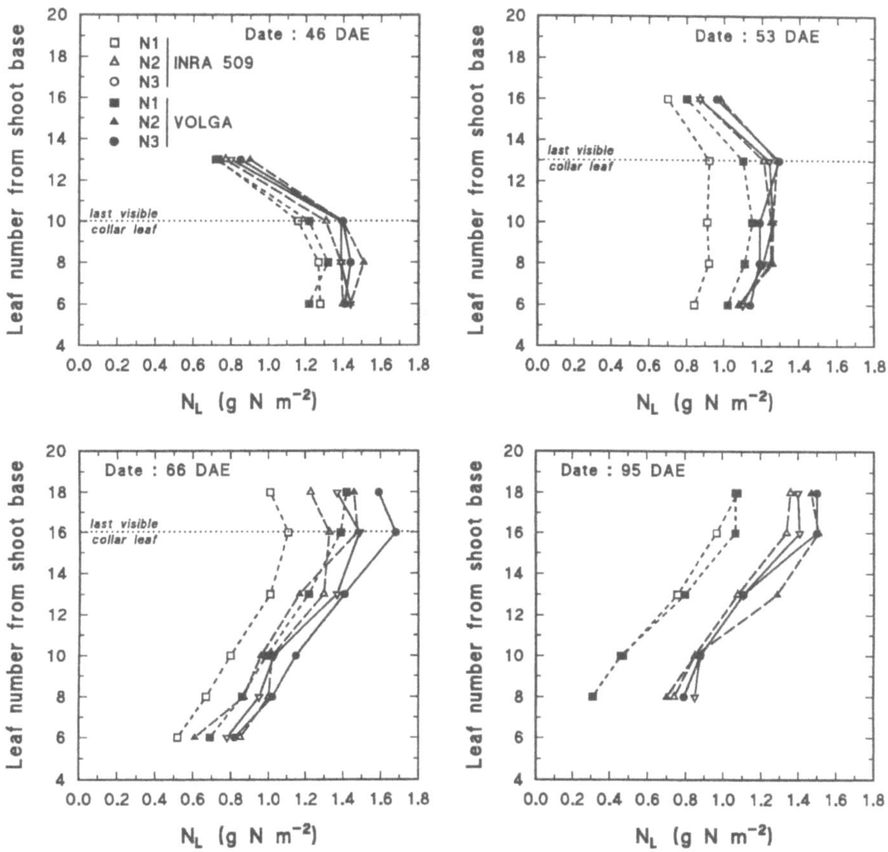

a b c d

Fig. 11.4a–d. Profiles of N per unit area (N_L) down the canopies of maize crops, cultivars INRA 509 and Volga, grown at Lusignan in 1994 with three N fertilizer application rates, *N1* 30 kg N ha^{-1}; *N2* 150 kg N ha^{-1}; *N3* 180 kg N ha^{-1}. Data for four sampling dates, expressed as days after emergence (DAE)

Nevertheless, the results obtained are encouraging, since the maximum N_L values at the canopy top seemed to discriminate well between treatments with optimum NNI and treatments with deficient NNI. This is illustrated in Fig. 11.6a and b, which show shoot N% and maximum N_L values for selected N treatments plotted together on the same graphs against accumulated shoot dry matter. It can be seen how, in comparison to shoot N%, maximum N_L changed very little as dry matter increased, while N% is only of value when the dry matter is known and reference is made to the critical dilution curve. Note the proportionately large difference in maximum N_L between N treatments at biomasses greater than 1000 g m^{-2}. This is at a stage when diagnosis of N status may be important for supplementary fertilizer applications but when N% can be less sensitive to differences in N supply, due to dry matter dilution.

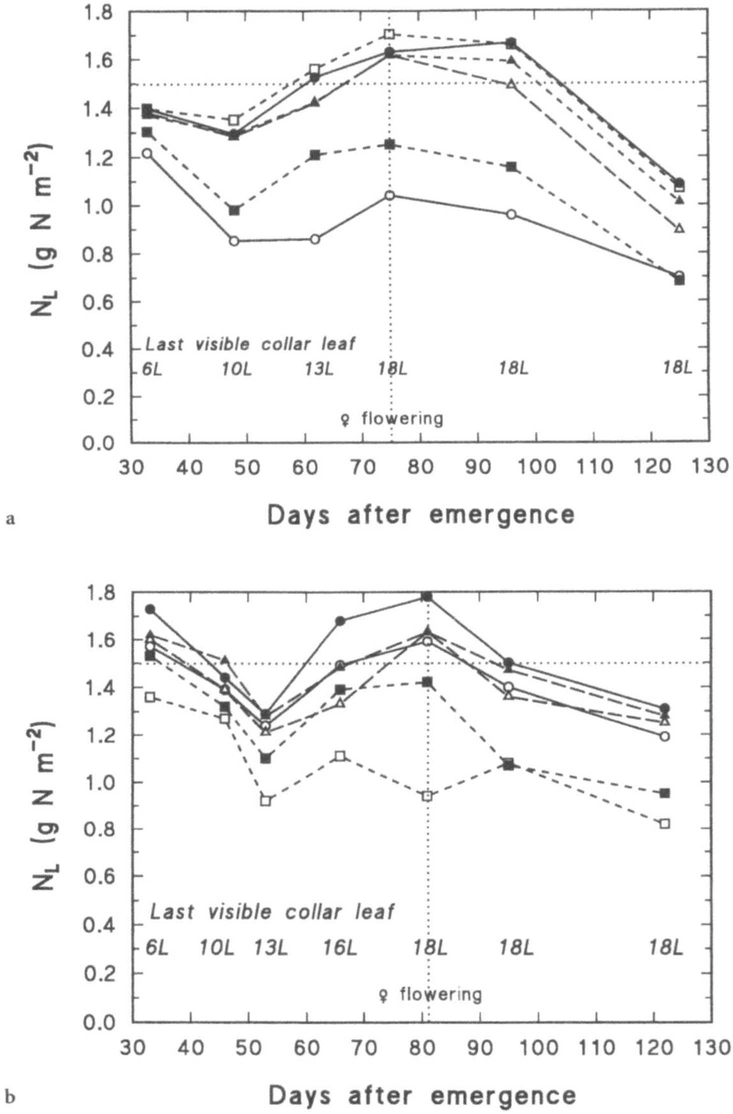

Fig. 11.5a,b. Time course of N per unit leaf area (N_L) in last visible collar leaves of maize grown at **a** Onard and **b** Lusignan. **a** Cultivar Volga with N applied at 50 (○), 90(■), 130 (△), 190 (▲), 240(●), and 290 kg N ha^{-1} (□). **b** Cultivars INRA 509 (*open symbols* and *dashed lines*) and Volga (*closed symbols* and *solid lines*) with N applied at 30 (□ ■), 150 (△ ▲), and 180 kg N ha^{-1} (○ ●)

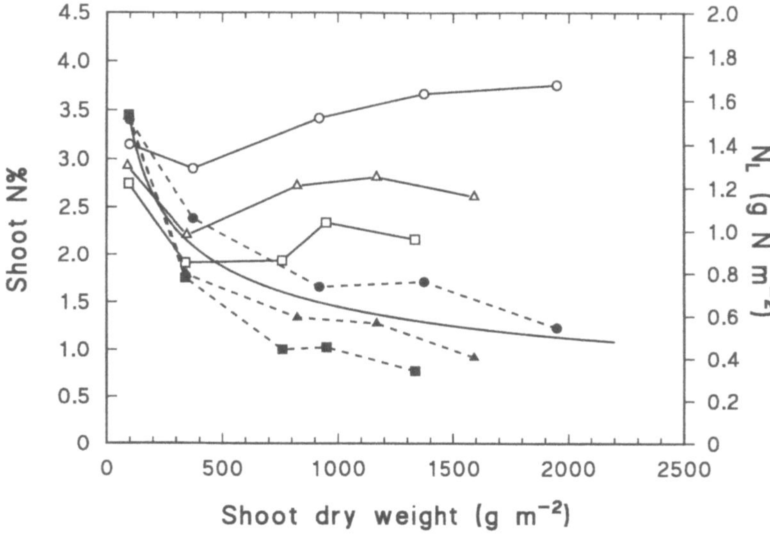

a

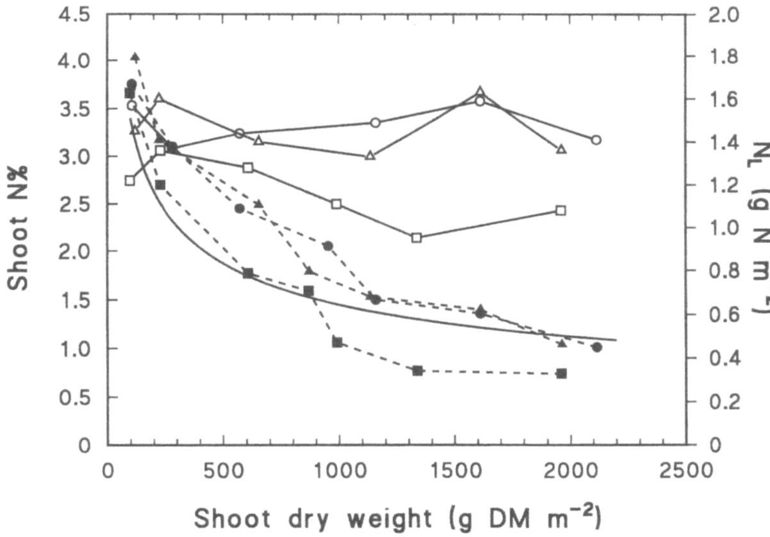

b

Fig. 11.6a,b. Shoot N% (*solid symbols* and *dashed lines*) and N per unit leaf area (N_L) in last visible collar leaves (*open symbols* and *solid lines*) plotted against shoot dry weight for maize grown at **a** Onard and **b** Lusignan. **a** Cultivar Volga with N applied at 50 (■ □), 90 (▲ △), and 240 kg N ha^{-1} (● ○). **b** Cultivar INRA 509 with N applied at 30 (■ □), 150 (▲ △), and 180 kg N ha^{-1} (● ○). *Solid curve* in each figure represents critical N%

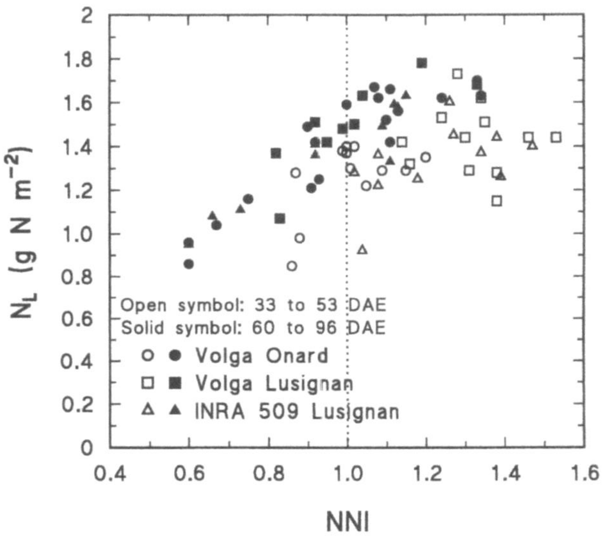

Fig. 11.7. N per unit leaf area (N_L) in last visible collar leaves plotted against nitrogen nutrition index (*NNI*) for maize grown at Onard and Lusignan in 1994. *Dotted line* shows NNI = 1, below which N supply is limiting

Finally, Fig. 11.7 shows maximum N_L plotted against NNI, combining data from both the sites. Allowing for the small variation in maximum N_L with time, and the fact that it was not always the last collar leaf that was sampled, there was a reasonably good correspondence between these two potential indicators of N nutrition status. As N supply became limiting and NNI decreased below 1, maximum N_L also decreased below the critical N_L of around 1.5 g m^{-2}. With excess N supplies, NNI tended to increase without a corresponding increase in the maximum N_L, which was presumably due to luxury consumption of N in tissues of the stem.

11.4 Conclusion

The use of a critical N_L, here defined as the leaf N per unit leaf area in the uppermost mature leaves of the canopy, appears to have potential for the diagnosis of N nutrition status in crops such as maize, and further research seems justified. A great advantage of this approach is the relative constancy of the critical N_L as dry matter accumulates, so that it is not necessary to measure biomass. A simple farm system can be envisaged where leaf disks are punched from the uppermost mature leaves and their N content is analyzed, to give a direct diagnosis of crop N status. Before this can be done, instantaneous values of critical N_L, and their variation throughout the growing season need to be determined more precisely.

References

Anten NPR, Schieving F, Werger MJA (1995) Patterns of light and nitrogen distribution in relation to whole-canopy carbon gain in C_3 and C_4 mono- and dicotyledenous species. Oecologia 101:504–513

Beaufils ER (1973) Diagnosis and recommendation integrated system (DRIS). Soil Science, Bull N°1, University of Natal, South Africa

Bever HB (1986) Foliar diagnosis of soybean by DRIS. Commun Soil Plant Anal 17:237–356

Charles-Edwards DA, Stutzel H, Ferraris R, Beech DF (1987) An analysis of spatial variation in the nitrogen content of leaves from different horizons within a canopy. Ann Bot 60:421–426

Edwall AM, Gascho GJ, Summer ME (1985) DRIS norms for 11 nutrients in corn leaves. Agron J 77:506–508

El Wali AMO, Gascho GJ (1984) Soil testing, foliar analysis, and DRIS as guide for sugar cane fertilization. Agron J 76:466–470

Evans JR (1989) Photosynthesis and nitrogen relationships in leaves of C_3 plants. Oecologia 78:9–19

Gastal F, Bélanger G, Lemaire G (1992) A model of the leaf extension rate of tall fescue in response to nitrogen and temperature. Ann Bot 70:437–442

Gonzalez-Montaner JH, Meynard J-M, Mary B (1987) Contrôle de la nutrition azotée du blé par l'analyse des teneurs en nitrates dans la plante. CR Acad Agric Fr (Paris) 73(3):105–115

Lagatu H, Maume L (1926) Diagnostic de l'alimentation d'un végétal par l'évolution chimique d'une feuille convenablement choisie. CR Acad Sci Fr 182:633–655

Lemaire G, Onillon B, Gosse G, Chartier M, Allirand JM (1991) Nitrogen distribution within a lucerne canopy during regrowth: relation with light distribution. Ann Bot 68:483–488

Meldal-Johnsen A, Summer ME (1980) Foliar diagnostic norms for potatoes. J Plant Nutr 2:569–576

Muchow RC, Sinclair TR (1994) Nitrogen response of leaf photosynthesis and canopy radiation use efficiency in field-grown maize and sorghum. Crop Sci 34:721–727

Pettygrove GS, Jackson LF, Sailsberg RL, Pelton RL, Hart J (1984) Determining nitrogen requirements for wheat using tissue analysis. *Agron Abstr*, American Society of Agronomy, p 217

Recous S, Fresneau C, Faurie G, Mary B (1988) The fate of labelled ^{15}N urea and ammonium nitrate applied to a winter wheat crop. I. Nitrogen transformation in the soil. Plant Soil 112:205–214

Sinclair TR, Horie T (1989) Leaf nitrogen, photosynthesis, and crop radiation use efficiency: a review. Crop Sci 29:90–98

Smith PF (1962) Mineral analysis of plant tissues. Annu Rev Plant Physiol 13:81–108

Talouzie A, Guiraud G, Moyse A, Marolle T, Champigny ML (1984) Effect of previous nitrate deprivation on ^{15}N-nitrate absorption and assimilation by wheat seedlings. J Plant Physiol 116:113–122

Ulrich A (1943) Plant analysis as a diagnostic procedure. Soil Sci 55:101–112

Walworth JL, Summer ME (1987) The diagnosis and recommendation integrated system (DRIS). Adv Soil Sci 6:149–188

Werger MJA, Hirose T (1991) Leaf nitrogen distribution and whole canopy photosynthetic carbon gain in herbaceous stands. Vegetatio 97:11–20

CHAPTER 12

Radiometric Estimates of Nitrogen Status of Leaves and Canopies

F. Baret and Th. Fourty

Contents

12.1 Introduction

Estimation of the nitrogen status of plants provides key information for optimum management of fertilizer applications to crops as reviewed in Chapters 9 and 11 (this Vol.). Most of the techniques currently used are based on plant sampling, generally using the biochemical composition of plant tissues and mainly nitrogen concentration. These techniques are generally tedious, time-consuming, destructive, and thus cannot not be repeated many times for a more representative evaluation of the canopy nitrogen status in a field or amongst fields of a given area.

Remote sensing techniques offer great opportunities to complement these classical sampling procedures by simple radiometric measurements. This can be achieved at the very local scale of the leaves by using field devices designed to

INRA Station de Bioclimatologie, Domaine Saint Paul, Site Agroparc, 84914 Avignon Cédex 9, France

Diagnosis of the Nitrogen Status in Crops
G. Lemaire (ed)

estimate, generally indirectly, the nitrogen concentration. Leaf reflectance or transmittance measurements are very easy to perform and thus allow a great improve in the sampling scheme. Radiometric measurements could also be performed from a few meters above the crop with an area of the order of a square meter being sampled instantaneously. The variation in wavelength of canopy reflectance is obviously governed by leaf optical properties. However, confounding factors such as the canopy structure, the illumination and viewing geometry, and soil optical properties may induce ambiguities in the interpretation. When using airborne or satellite-borne data, the atmosphere will, in addition, induce some perturbations in the light reflected by the canopy as observed from the space platform. However, these systems potentially provide an exhaustive coverage of a given region, allowing an instantaneous estimate of the status of all the fields. Furthermore, a monitoring is potentially accessible from frequent observations throughout the growing season. The objective of this chapter is to review the current techniques that have been developed to infer the nitrogen status of plants from radiometric measurements either at the leaf, canopy, or space levels. We will first address the leaf level before tackling the canopy level. We will eventually conclude about the potential applications at the space level.

12.2 Estimating Nitrogen Concentration at the Leaf Level

12.2.1 Measurements of Leaf Optical Properties and Physical Processes

Leaf optical properties are characterized by reflectance, transmittance, and absorptance. These quantities correspond to the fraction of incident radiation that is respectively reflected, transmitted, or absorbed by a leaf. Two main processes thus govern radiative transfer in the leaf: scattering and absorption.

1. Scattering. Scattering is mainly due to discontinuities in the refractive index encountered at the interfaces between air, cell walls and cytoplasmic water. The spectral variation of optical properties due to scattering is not very important since it is governed by the spectral variation of the refraction index which does not vary much with the wavelength (Fig. 12.1).
2. Absorption. Elementary processes of absorption are due either to electronic transitions or biochemical bond vibration, rotation and stretching. Electronic transitions occur at the lower wavelengths, mainly in the visible spectral domain, up to 800 nm. Pigments such as chlorophyll, carotenoids, or xanthophyll absorb this way. Biochemical bonds between light atoms such as C, N, O, or H absorb at specific frequencies and their harmonics corresponding to vibration, rotation, or stretching of the bonds. These processes occur mainly in the short wave infrared domain, above 800 nm. Thus, these physical principles show that no direct estimate of N content is possible. However, N in leaves is bound with C, H, or O atoms,

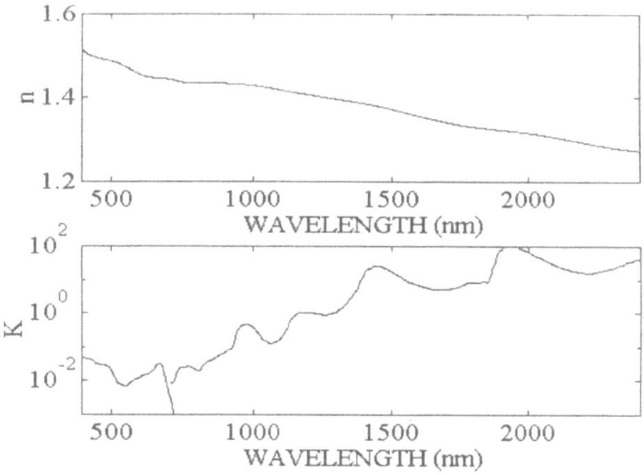

Fig. 12.1. Spectral variation of the refraction index (Jacquemoud and Baret 1990)

mainly within protein molecules. This implies that is should be possible to infer N content from the absorption features of protein, or from the constituents that have characteristic absorption features and are strongly correlated to the N content per unit leaf area. Note that here we consider only N content on a leaf area basis and not N concentration. Concentration corresponds to the mass of absorbing materials per unit leaf dry mass. The specific leaf weight (SLW), that is the leaf dry mass per unit leaf area, is used to convert from concentration (C_N), to content (N%) to

$$N\% = N_L/SLW.$$

The light that is reflected or transmitted by a leaf might possibly interact within the whole leaf thickness in spectral domains of low absorption. Reflectance and transmittance are thus sensitive to variation in the content of absorbing constituents per unit leaf area.

Because intrinsic scattering properties are wavelength-dependent to only a small extent, the spectral variation of leaf optical properties is mainly due to the absorption characteristics of leaf constituents. The determination of leaf N content requires us to identify particular absorption features for protein or other constituents such as chlorophyll, to relate protein or chlorophyll content to N content, and then to derive N concentration using the specific leaf weight. This general approach is shown diagramatically in Fig. 12.2.

We will attempt an analytical approach first, using a simple radiative transfer model to identify protein absorption features. This should provide more robustness and reliability for the relationships developed. In parallel, we will investigate chlorophyll and dry mass absorption features. This will allow us to evaluate the possibilities of obtaining protein and chlorophyll contents as well as the specific leaf weight. We will then look for alternative approaches using empirical relationships as a short-cut relationship between optical properties and N concentration,

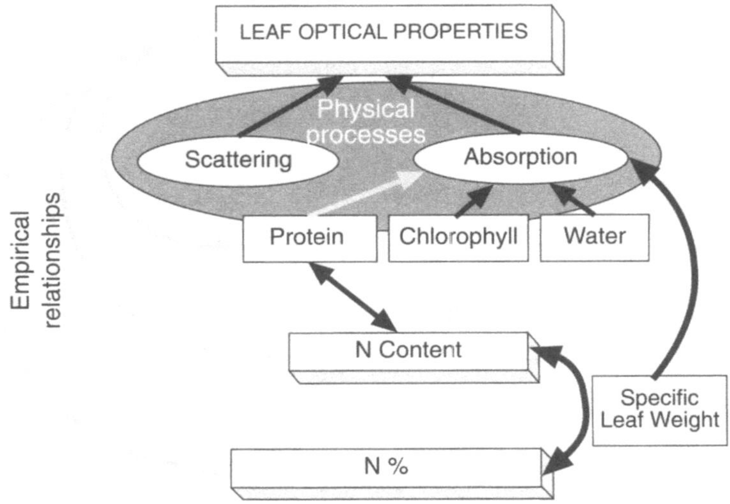

Fig. 12.2. Schematic description of several approaches used to derive leaf nitrogen concentration (*N%*) from optical leaf properties. *Black arrows* correspond to strong and robust relationships. *Gray arrows* correspond to empirical relationships that may not be as reliable and generalizable

or using more indirect relationships such as those possibly existing between chlorophyll and N contents.

12.2.2
The Analytical Approach

Based on the physical principles described in brief earlier, we will evaluate the possibility of determining N concentration. We will first describe a simple radiative transfer model that mimics the interaction of light within leaves, namely, the PROSPECT model (Jacquemoud and Baret 1990).

12.2.2.1
The PROSPECT Leaf Optical Properties Model

Light transfer within the leaf can be approximated by a pile of N identical layers characterized by their refraction index (n), and global absorption coefficient K (Fig. 12.3). This global absorption coefficient depends on the content of absorbing molecules in the leaf such as chlorophyll, water, protein, etc.

$$K = \sum_i k_i C_i ,$$

where C_i and k_i are, respectively, the content per unit leaf area and the specific

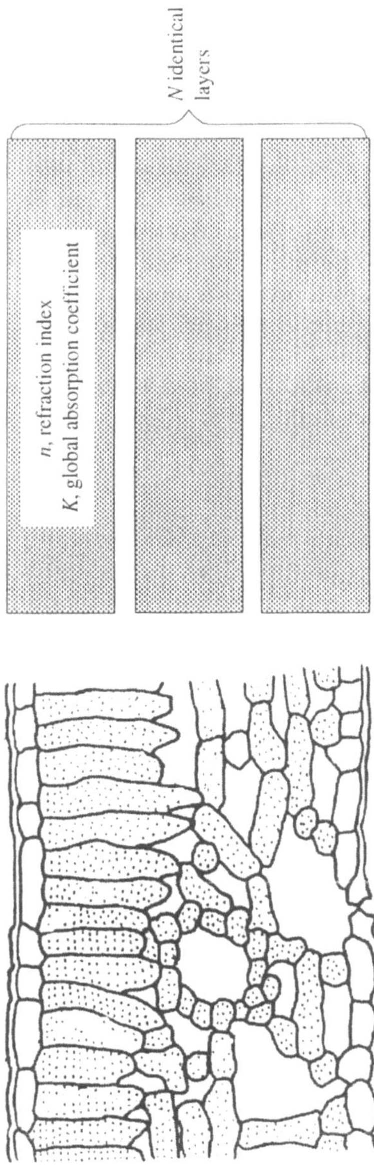

Fig. 12.3. *Left* Typical structure of a real leaf; *right* leaf structure as used in the PROSPECT model

absorption coefficient of constituent i. Specific absorption coefficients represent the absorption efficiency of a unit mass of constituent per unit leaf area. They have the unit of the reciprocal of the content per unit leaf area. They depend on the constituent considered and to a great extent on the wavelength.

Figure 12.4 shows the spectral variation of the two main absorbing components of leaves, chlorophyll and water. We will see later that it is possible to derive a specific absorption coefficient for protein. The PROSPECT model is quite simple and can be inverted to retrieve the biophysical and biochemical characteristics of the leaf from the spectral variation of reflectance and/or transmittance. The model inversion process consists of looking for the best set of leaf characteristics (such as contents, specific absorption coefficients, leaf structure parameters) that provides simulated reflectance and transmittance values as close as possible to the observed one. Model inversion will be used to retrieve protein absorption characteristics.

12.2.2.2
Characterizing Protein Absorption Features

Fourty et al. (1996) derived a specific absorption coefficient for protein using the PROSPECT model applied to an extensive leaf data set (Hosgood et al. 1995) for which both optical properties and biochemical composition were determined.

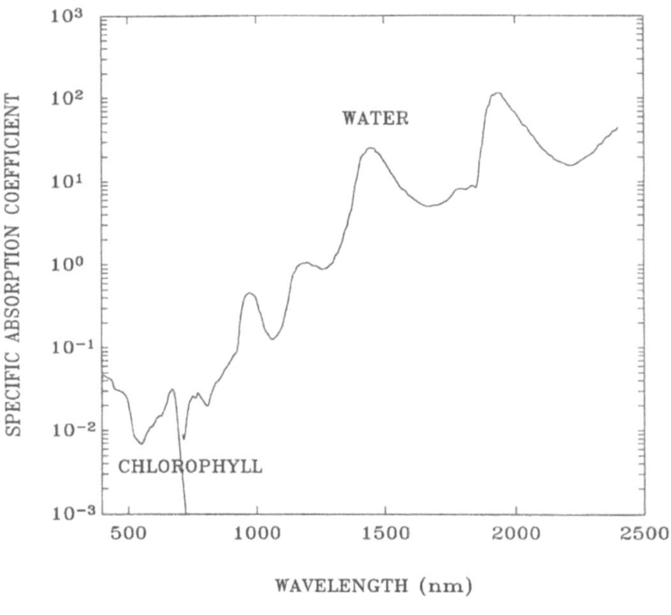

Fig. 12.4. Spectral variation of the specific absorption coefficients (K) of chlorophyll and water

Figure 12.5 shows the estimated specific absorption coefficient for protein. We note that most of the absorption bands reported by Curran (1989) in his review of remote sensing of canopy biochemistry correspond to absorption peaks of the estimated specific absorption coefficient for protein.

There results show that the absorption features of protein are quite well characterised. However, they are not very strong as compared to the absorption features of other constituents that absorb in the same spectral domain (1300–2500 nm), such as water or leaf dry matter (the latter includes protein, but also cellulose, hemicellulose, sugar, lignin, starch, etc.). Figure 12.6 shows that the contribution of protein to the leaf absorption is indeed very small. This implies that we first have to evaluate the sensitivity of leaf optical properties to protein content, before investigating the possibility of directly determining the latter.

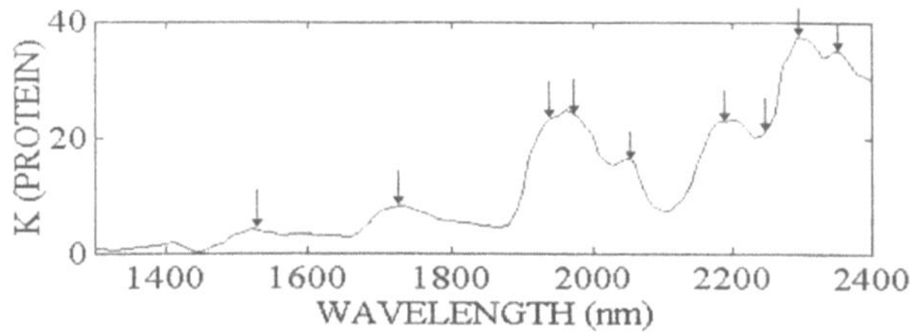

Fig. 12.5. Specific absorption coefficient for protein. *Arrows* show the main absorption features reported by Curran (1989)

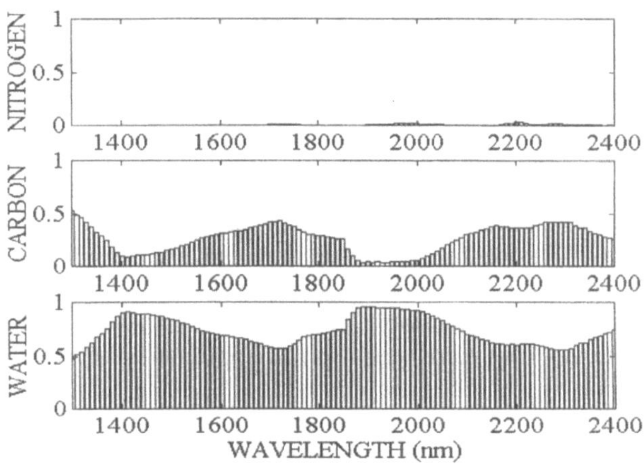

Fig. 12.6. Average contribution of protein, water and dry matter to the leaf absorption. The contribution for constituent i is computed as: $\overline{C}_i k_i / \Sigma_j \overline{C}_j k_j$ where $\overline{C}_i$ represents the average content of constituent i

12.2.2.3
Protein Content Retrieved from Model Inversion

Sensitivity Analysis. The leaf optical properties model described previously allows a simulation of leaf reflectance and transmittance spectra as a function of protein content. It will be used to first analyze the sensitivity of leaf optical properties to protein content. To simplify the investigation, we will consider the main absorbers within the leaf as being made of water, carbon, and nitrogen in the 1300–2400-nm range. The analysis will be carried out using the previously determined values of the specific absorption coefficients of protein, and that for dry matter provided by Fourty et al. (1996). For water, we used the values proposed by Curcio and Petty (1951) for pure water. Protein content is varied for a generic leaf defined by the following variables:

Water content:	0.01 cm.
Specific leaf weight:	0.006 g cm^{-2}.
Leaf mesophyll structure index (N):	1.5.

The protein content is varied from 6 to 36%, the average content being close to 18%.

The results show that both reflectance and transmittance express little sensitivity to changes in nitrogen content (Fig. 12.7). Note that the transmittance is slightly more sensitive to the protein content than is reflectance. In the best case, corresponding to the transmittance in strong protein absorption bands close to 2200 nm, the amplitude of variation of transmittance is smaller than 0.005. This implies that an accurate estimate of protein content will be possible only with very accurate leaf optical properties measurements, assuming that the other confounding factors such as variation in leaf structure and water content can be effectively accounted for. In the following, we will evaluate the accuracy of protein content retrieval from leaf optical properties measurements.

N Content Retrieval from Model Inversion. Fourty et al. (1996) attempted to retrieve protein content from measurements of optical properties in dry leaf by inverting the PROSPECT model using the previously determined specific absorption coefficients. The use of dry rather than fresh leaves should facilitate protein content retrieval, since water absorption features will no longer mask those of protein. However, these authors found great difficulties in accurately estimating protein content. They concluded that the estimation of protein content from dry leaf spectra is not accurate enough to be relevant. These poor results are mainly explained by the weakness of the absorption signal associated with protein. Nevertheless, these results are actually restricted to the retrieval of protein content using a physical model. The model may not be perfect, and possible errors could come from certain assumptions and approximations. One important assumption is that protein absorption features are stable and do not depend on the type of leaf

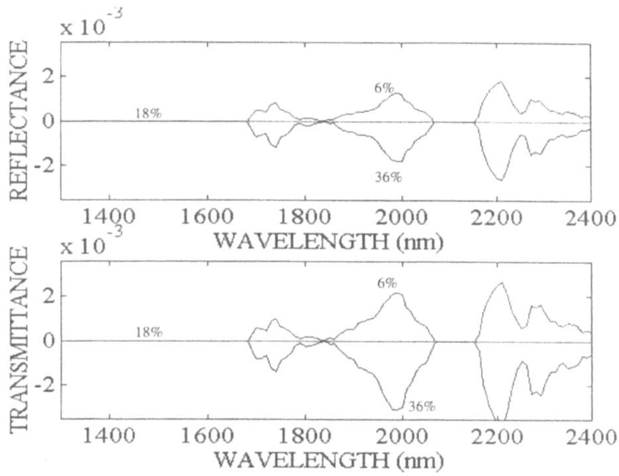

Fig. 12.7. Sensitivity of leaf reflectance and transmittance to variations in protein content. The spectra presented correspond to the difference from the standard case (18%) spectra

considered. The absorption features depend on the composition and steric arrangement of the molecules, leading to some scattering in the absorption features as a function of the abundance of the various protein types. This would correspond to a lack of specificity in protein absorption features. For this reason, we will also evaluate a classical empirical approach that should implicitly take into account all these problems through the covariance of protein content with other variables that can be sensed by a spectrophotometer.

12.2.3
Empirical Approaches

Here we will investigate the possibility of retrieving nitrogen content per unit area indirectly by estimating the content of other absorbers that may be closely related to the nitrogen content. However, before this, we will evaluate the performances of a purely empirical relationship between N content and leaf optical properties.

12.2.3.1
The Empirical Relationship Between N Content and Leaf Optical Properties

In the classical techniques of near-infrared spectrophotometry, the determination of leaf biochemical composition is routinely performed using the stepwise regression (Williams and Norris 1987; Marten et al. 1989; Dardenne 1990). This statistical tool is used to relate the content of biochemical molecules to the reflectance spectra, generally measured in ground dry leaf material using a spectrophotometer. Due to the generally nonlinear pattern of the relationship between reflectance and content, reflectance (ρ) values are transformed into absorbance (α) according to: $\alpha = \log_{10}(1/\rho)$.

Empirical approaches are generally restricted to the conditions that prevailed during the experiment. It is thus critical to evaluate the robustness of the relationships by testing their predictive performances over independent data sets. This was done in the study described in the following, which used the LOPEX data set (Hosgood et al. 1995). This consists of a collection of 58 very different leaves, for which the detailed biochemical composition, including nitrogen and chlorophyll content, was measured along with leaf reflectance and transmittance spectra. The main factors leading to variation from one leaf to another were the species, light exposure, nutrient level, and phenological stage. The data set was first randomly split into 2 subsets of 29 individuals each. One was used to calibrate the relationships (calibration data set). The other was used to evaluate the predictive performances of the previously calibrated relationships (test data set). The results (Fig. 12.8) show that when calibrating and testing the relationship on the same data set, we obtain a very good fit between estimated and measured values of protein content. The goodness of the fit increases as a function of the number of wavelengths selected. According to the observations of Hruschka (1987), a ratio between 5 to 10 samples per wavelength selected should be used to prevent overfitting problems. For our data set having 29 samples, this was observed in practice when evaluating the predictive performances of the previously calibrated relationships on the test data set (Fig. 12.8). Maximum R^2 values were obtained for three to four wavebands. Adding more wavebands decreased the predictive performances of the relationship. However, the maximum R^2 values obtained were not very high and corresponded to inaccurate estimates of protein content.

These results reveal the lack of robustness of such empirical methods, particularly when applied to small data sets. This is in good agreement with the conclusions of Grossman et al. (1995), who found that empirical relationships applied to several independent experiments were leading to very poor predictive perfor-

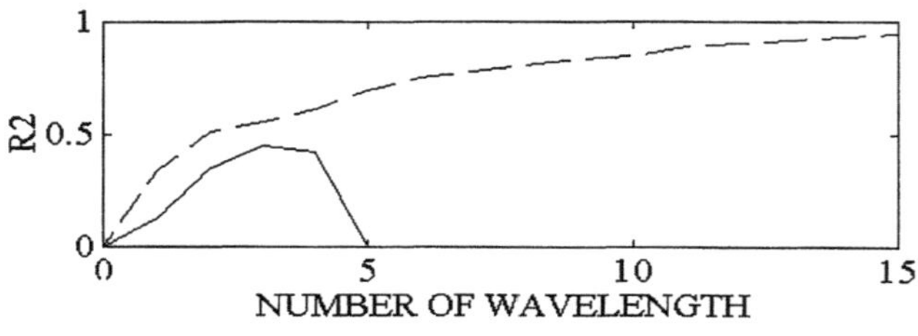

Fig. 12.8. Values of the determination coefficient (R^2) for regressions of measured values of protein content on estimated values as a function of the number of wavelengths selected. The *solid line* corresponds to the case where the relationship established on the calibration data set is applied to the test data set. The *dashed line* corresponds to the case where the relationship established on the calibration data set is evaluated over the same calibration data set

mances. It also confirms our previous results obtained using model inversion: the weakness of the protein absorption signal and its poor specificity prevent us from accurately and reliably estimating protein content. This apparent contradiction with near-infrared spectroscopy as routinely used is explained by the fact that in near-infrared spectroscopy, the empirical relationships are generally calibrated and used over leaves from one species, and sometimes one cultivar. Furthermore, they are generally developed using dry ground materials which have no water to mask the absorption features of other compounds, and which standardize the scattering due to leaf structure.

As a preliminary conclusion, it seems quite difficult to estimate reliable nitrogen or protein contents of fresh leaves from their optical properties. However, we still have to investigate the possibility of estimating leaf nitrogen content from the determination of the content of the main absorbers. Chlorophyll pigments are interesting candidates because of the strength of their absorption features. The other candidate could be the specific leaf weight.

12.2.3.2
Estimation of N Content from Chlorophyll Content and Specific Leaf Weight

The estimation of N content from chlorophyll content and specific leaf weight measurements obviously requires the relationships between N content and chlorophyll or specific leaf weight to be strong. We therefore have to evaluate these relationships first of all.

Relationship Between N Content, Chlorophyll Content and Specific Leaf Weight. In this section we will consider chlorophyll content without distinction between the a and b forms. These two forms are generally very strongly correlated, so that it is almost impossible to obtain independent estimates of the content in chlorophyll a and b that could be actually interpretable. Ercoli et al. (1993) reported a strong correlation between chlorophyll and nitrogen contents on a leaf area basis. This was observed in maize grown under a range of N supplies. The oldest green leaves and most recently fully expanded leaves were sampled at two growth stages. Figure 12.9 shows that chlorophyll content was quite strongly and positively correlated to nitrogen content. However, using the LOPEX data set (Hosgood et al. 1995) which was described earlier, we have observed a very poor but positive relationship between chlorophyll and nitrogen contents (Fig. 12.9). Furthermore, LOPEX and Ercoli's data sets are not in very good agreement, with an apparent bias. This implies that there is no general relationship between nitrogen and chlorophyll contents. The processes that would sustain such a relationship are complex as reviewed by Evans (1989) and mainly depend on the partitioning between thylakoids and soluble proteins. The estimation of nitrogen content from chlorophyll content requires to develop and apply the relationship under very well controlled and delimited conditions such as those described by Ercoli et al. (1993).

When investigating the relationship between N content and specific leaf

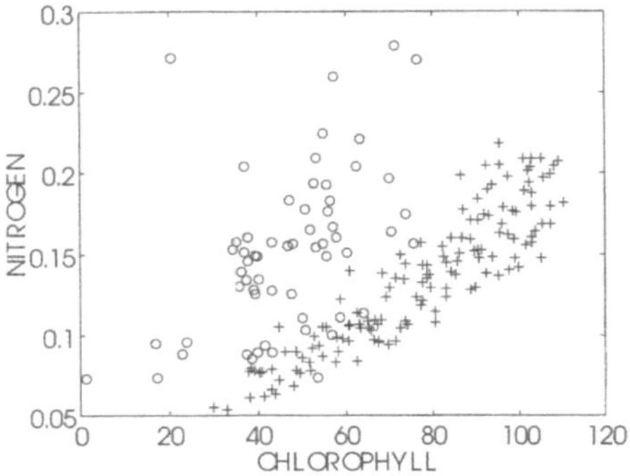

Fig. 12.9. Relationship between chlorophyll and nitrogen contents per unit leaf area (mg cm^{-2}). The *circles* and *crosses* correspond, respectively, to the LOPEX (Hosgood et al. 1995) and Ercoli et al. (1993) data sets

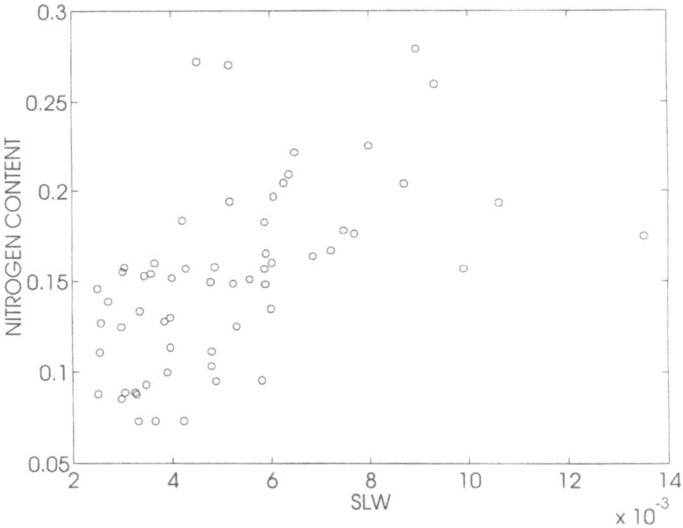

Fig. 12.10. Relationship between specific leaf weight (g cm^{-2}) and nitrogen content per unit leaf area (mg cm^{-2}). Data from the LOPEX experiment (Hosgood et al. 1995)

weight (Fig. 12.10), a positive correlation appears which is not surprising: the more dry mass the leaf has, the more protein, and thus N, would be expected in the leaf. However, the correlation is not very strong, and a considerable degree of scattering is observed around the main trend, leading to poor predictive performances.

The relationship between N concentration and specific leaf weight is more interesting. Field and Mooney (1986) report a significant relationship between specific leaf weight and leaf nitrogen concentration. The LOPEX data set shows a similar trend (Fig. 12.11). For small values of specific leaf weight (SLW <

0.010 g cm^{-2}), generally corresponding to young compact leaves, N concentrations vary widely. This should result mainly from the variation of the N concentration in metabolic leaf material. Conversely, for larger SLW values (SLW > 0.010 g cm^{-2}), N concentration remains more or less stable, close to a value of 2% that should be mainly related to the N concentration of structural leaf material. Therefore, the SLW is a very poor indicator of nitrogen concentration for SLW values below 0.01.

These results indicate the difficulty of setting up a general relationship between chlorophyll content or specific leaf weight, and nitrogen concentration or content. However, it seems that for restricted experimental conditions such as the ones that prevail for the management of N fertilization of a particular crop, it is possible to use empirical relationships between N concentration and chlorophyll content or specific leaf weight. As an example, Turner and Jund (1991) demonstrate in rice crops that it is possible to determine rice N requirements from leaf chlorophyll content estimates at certain growth stages. In the following, we will describe the methods used at the leaf level to retrieve chlorophyll content and specific leaf weight from measurements of the spectral variation in their reflectance or transmittance.

Estimation of Chlorophyll Content. The estimation of chlorophyll concentration from leaf reflectance and/or transmittance measurements has been proposed over many years by several authors (Benedict and Swilder 1961; Hardwick and Baker 1973; Wallihan 1973; Macnicol et al. 1976; Hardacre et al. 1984; Yadava 1986; Spomer et al. 1988; Peng et al. 1993; de Rosny et al. 1995). All the methods are based on the use of the spectral variation of chlorophyll absorption features (Fig. 12.12). Most of the studies are based on transmittance rather than reflectance measurements. Transmittance is preferred because it is more sensitive than reflectance to chlorophyll concentration, and less sensitive to confounding factors such as the surface reflec-

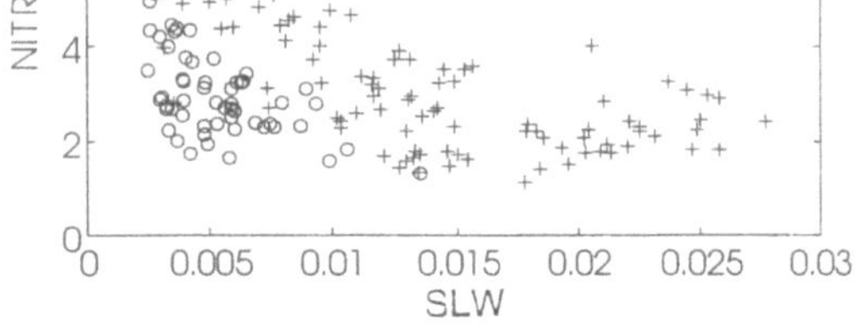

Fig. 12.11. Relationship between leaf specific weight (g cm^{-2}) and leaf nitrogen concentration (N%). The *circles* and *crosses* correspond respectively to the LOPEX (Hosgood et al. 1995) and Field and Mooney (1986) data sets

tive properties. Transmittance is, in this case, generally determined under normal illumination and viewing measurement configurations.

Leaf transmittance, τ, in a given waveband λ, can be approximated as a simple function of chlorophyll concentration according to Baret et al. (1988) as follows:

$$\tau(\lambda) = \tau_0 e^{-K(\lambda).C_{ab}},$$

where $K(\lambda)$ is an absorption coefficient, C_{ab} the chlorophyll content per unit leaf area, and τ_0 the transmittance of the leaf if it was deprived of chlorophyll ($C_{ab} = 0$). When no other compound except chlorophyll absorbs, the maximum transmittance τ_0 depends only slightly on the wavelength because the scattering properties in a limited spectral domain can be considered as spectrally independent, as seen earlier. Thus, the value τ_0 is well approximated by the transmittance measured in the near-infrared. It follows that the ratio between the transmittance in a wavelength where chlorophyll absorbs to that observed in the near-infrared (τ_0) where the absorption is minimum, is simply related to the chlorophyll content:

$$\frac{\tau(\lambda)}{\tau_0} = e^{-K(\lambda).C_{ab}}.$$

More recently, the Minolta Company (Minolta Camera Co. 1991) has built a device, the SPAD-502 chlorophyllmeter, that allows rapid in situ estimates of chlorophyll content from the leaf transmittance measured in two wavebands (650 and 940 nm). Figure 12.12 shows some results obtained using this method on a range of data

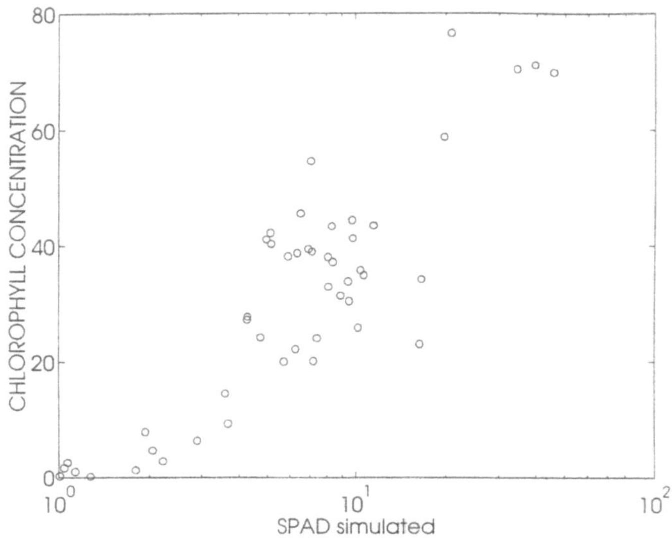

Fig. 12.12. Relationship between the SPAD output [Log(T940/T650) = $\text{Log}(\tau_0/\tau(\lambda = 650\,\text{nm}))$] with $\tau_0 = \tau(\lambda = 940\,\text{nm})$ and the measured chlorophyll a and b content (mg cm^{-2}). The RMSE is 9.7 mg cm^{-2} and the R^2 is 0.758. Data measured over a wide range of leaves (Penuelas et al. 1994)

presented in Penuelas et al. (1994). We observe a quite strong correlation. However, some significant errors appear, mainly due to the choice of the wavelengths and the method used to correct for the effect of leaf structural effects. Detailed simulations with the leaf radiative transfer model show that this way of estimating chlorophyll content from the ratio between the near infrared and red transmittance leads to some inaccuracies. In particular, the choice of a band centered close to the maximum chlorophyll absorption at 680 nm is not optimal for higher chlorophyll concentration values because of the very small sensitivity of the transmittance in those conditions. The choice of two or three bands in the red edge in between 680 and 780 nm would allow the method to be more sensitive for the widest range of situations. Improvement of the method is under investigation.

Estimation of Leaf Specific Weight. Very few works report estimation of the specific leaf weight from leaf optical properties measurements. Baret and Fourty (1996) demonstrated good estimates of SLW obtained using the PROSPECT model inversion over high spectral resolution measurements of the reflectance or transmittance of fresh leaves. However, they observed that using the transmittance rather than the reflectance led to better results (Fig. 12.13). Empirical approaches based on stepwise linear regression were also investigated, and led to similar results (Fourty and Baret 1996): accurate estimates of SLW could be obtained in preference by using the transmittance values. In this case, three to six wavebands located in the 1400–2400 nm range are required.

The review here of techniques used to estimate N content shows that it is obviously not possible to derive general relationships for the estimation of nitrogen concentration from the spectral variation of fresh leaf reflectance or transmittance. Thus, the only possibility that exists is to develop indirect relationships with N%. We have shown that the specific leaf weight and particularly chlorophyll

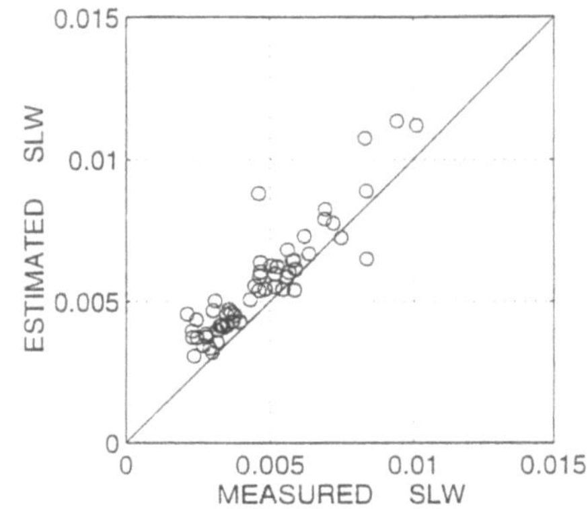

Fig. 12.13. Comparison between the measured *SLW* and that retrieved from model inversion using fresh leaf transmittance spectra measurements over the 1300–2400-nm domain ($R^2 = 0.793$, RMSE = 0.0010, n = 58)

content could be correlated to N content or N% under restricted experimental conditions. We demonstrated that chlorophyll content and SLW could be estimated with quite good accuracy from the spectral variation of the transmittance measured over fresh leaves. The design of the device that should measure the chlorophyll content and the SLW could be improved, based on the knowledge accumulated recently.

Another approach that could be investigated is the development of classical empirical regressions using near-infrared spectroscopy to predict N% in fresh leaves (Dardenne et al. 1991). However, this approach requires a definition of the domain of validity of the calibration equations (species, cultivar, leaf order, phenological stage, etc.) and the use of quite expensive devices capable of very accurately measuring the reflectance or transmittance of fresh leaves. However, this will certainly limit the potential of this methodology.

In the following section, we will investigate how the leaf content in the main absorbers could be estimated from reflectance measurements at the canopy level. Unfortunately, at this time, no studies report results on canopy level estimates of SLW. We therefore will focus only on the determination of leaf chlorophyll content from canopy level reflectance measurements.

12.3 Estimation of Chlorophyll Content at Canopy Level

In a first step, we will describe the sensitivity of canopy reflectance to leaf chlorophyll content and also possible confounding factors such as leaf or canopy structure. In the second step, we will present current developments aimed at estimating leaf chlorophyll content from canopy reflectance measurements.

12.3.1 Canopy Reflectance Sensitivity to Leaf Chlorophyll Concentration

Canopy reflectance depends on four sets of variables:

1. Leaf optical properties, that is leaf reflectance and transmittance. Thus chlorophyll content (C_{ab}), water content (C_w), SLW, and leaf mesophyll structure index (N) are the main variables that directly determine optical leaf properties, as seen in the previous sections.
2. Soil reflectance (ρ_s). This varies mainly with the soil type, moisture, and roughness. However, for a given area, soil type is fixed, and the only unknowns are roughness and moisture.
3. Canopy structure which is mainly characterized by the area, size, position and orientation of the plant elements such as leaves, stems, fruits, etc. For homogeneous crops, canopy structure could be represented as a first-order approximation by the leaf area index (l: area of leaves per unit soil area) and the leaf angle distribution (θ_l).

4. Measurement configuration, defined by the view (θ_o) and sun beam directions (θ_s).

It obviously follows that canopy reflectance results from the interaction of several variables. Radiative transfer models have been developed to describe the physical processes of photon-vegetation interactions. Among the simplest approaches, the SAIL model (Verhoef 1984, 1985) combined with the PROSPECT model simulates canopy reflectance with the following input variables: C_{ab}, C_w, SLW, N, l, θ_l, θ_o, θ_s, ρ_s. We will present only results on the main variables, i.e., l, θ_l, N, C_{ab}, and C_w, the other variables being set to standard values: nadir viewing and 45° sun zenith angle; soil reflectance set to 0.20 independent of the wavelength, for simplification. Note that we did not simulate the effect of SLW variations because these induce only minor changes in the canopy level reflectance. Computations were performed from 400 to 2450 nm with 2-nm steps. Figure 12.14 shows the simulations which can be summarized as follows:

- Leaf area index (l) is obviously the main variable affecting canopy reflectance over the whole spectrum (Fig. 12.14a). Generally, canopy reflectance increases when leaf reflectance is greater than soil reflectance. This is particularly true in the near-infrared domain. Conversely, in spectral domains characterized by strong absorption features, canopy reflectance decreases when leaf area index increases.
- Leaf angle inclination (θ_l) acts very similarly to leaf area index (Fig. 12.14b). Confusion between these two canopy structure variables is expected when retrieving canopy characteristics from canopy reflectance spectral variations if only the spectral variation is used.
- Leaf mesophyll structure index (N) induces small variations, mostly at wavelengths where leaves absorb very little, such as in the near infrared plateau (Fig. 12.14c). Note that, here again, N and θ_l have similar effects on canopy reflectance.
- Chlorophyll content induces large variations in canopy reflectance in the visible domain, particularly for low chlorophyll content (Fig. 12.14d). For higher values, saturation occurs in the strong chlorophyll absorption regions centered in the red (680 nm) and in the blue (400 nm). In a manner similar to leaves, the sharpest variation of reflectance values with regard to wavelength is located in the red edge, between 680 and 800 nm.
- Water content effects span over the near- and middle-infrared domains (Fig. 12.14d). Saturation effects similar to those observed for chlorophyll are observed for high water contents values in strong water absorption bands (1450 and 1950 nm).

In this analysis, we have shown that many factors could confound the effect of variation in chlorophyll content, particularly if only one waveband is considered. A large variation in sensitivity of canopy reflectance to canopy structure and leaf chlorophyll and water content is observed over the wavelengths studied. This differential sensitivity according to wavelength is one of the most interesting properties provided by spectral data. In the following, the techniques used to

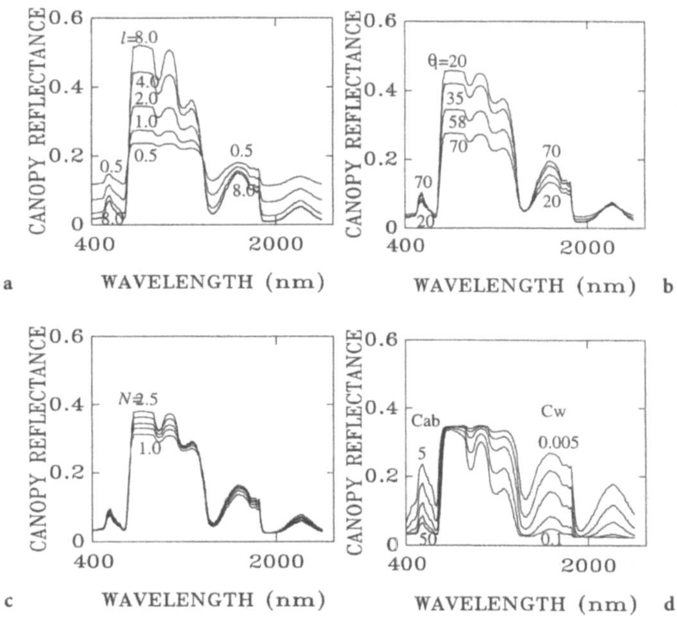

Fig. 12.14a–d. Sensitivity of canopy spectral reflectance to canopy characteristics. Variations around a standard canopy are considered. This standard canopy is characterized by: l = 2.0, θ_l = 58°, C_{ab} = 35 μg cm^{-2}, C_w = 0.02 cm, θ_0 = 0°, ϕ_0 = 0°, θ_s = 45°. The soil is assumed Lambertian with a constant reflectance value of 0.2. **a** Variations with leaf area index (l = 0.5, 1, 2, 4, 8). **b** Presents variations with leaf inclination (θ_l = 20, 35, 58, 70). **c** Variations with leaf mesophyll structure index (N = 1.0, 1.25, 1.5, 2.0, 2.5). **d** Simultaneous variations of leaf chlorophyll concentration, which affects the visible part of the spectrum (C_{ab} = 5, 10, 20, 35, 50 μg cm^{-2}), and leaf water equivalent thickness, which affects the near- and middle-infrared parts of the spectrum (C_w = 0.005, 0.01, 0.02, 0.05, 0.10 cm)

retrieve chlorophyll content will be based on this property to enhance the sensitivity to chlorophyll content and minimize that of the confounding factors.

12.3.2 Empirical Relationships

Empirical relationships have been proposed since the early 1980s to detect variations of vegetation reflectance due to chlorophyll concentration. They can be designed using either narrow spectral bands situated in the red edge or a small set of broad bands.

12.3.2.1 Empirical Relationships Based on the Red Edge

As seen earlier, the red edge corresponds to the sharp decrease of chlorophyll absorption from maximum absorption around 680 nm to almost zero absorption at 800 nm. In the 680–800-nm domain, soil reflectance and atmospheric effects (except for the oxygen absorption band) vary almost linearly and very little with the wavelength. Therefore, Baret et al. (1992) showed that the use of first or second derivatives of the reflectance versus wavelength provides a way to minimize the effect of these two factors. These authors demonstrated that the second derivatives could also be measured by wavelength shifts of the red edge. Several authors use shifts of the red edge or derivatives to characterize the vigor, chlorosis, or stress status of canopies (Chang and Collins 1983; Collins et al. 1983; Badhwar 1984; Miller et al. 1985, 1990; Baret et al. 1987; Demetriades-Shah and Steven 1988; Milton and Mouat 1989; Boochs et al. 1990; Demetriades-Shah et al. 1990; Rock et al. 1990). However, these studies do not explicitly relate the spectral characteristics of the canopy to the leaf chlorophyll concentration. The technique has mainly been used as a tool to qualitatively determine the physiological status of vegetation. Baret et al. (1992) showed from model simulations that the red edge provides composite information on both canopy structure and chlorophyll content. However, it seems quite difficult to decouple these two effects to identify the leaf chlorophyll content from canopy level measurements in the red edge. The red edge characterization can be achieved using only narrow spectral bands around 10 nm width. Other approaches based on broader bands have also been proposed.

12.3.2.2
Empirical Approaches Using Few Broad Bands

As we observed from the sensitivity analysis, leaf chlorophyll content variations influence canopy reflectance in the visible domain. However, this domain is also quite sensitive to canopy structure. Experimental results show that for a given species and cultivar, the leaf water content per unit area of green leaves is quite stable (Baret 1994). Therefore, canopy reflectance in the middle-infrared and near-infrared will mainly be determined by canopy structure and optical soil properties. Baret and Andrieu (1994) proposed to use concurrently classical vegetation indices computed in the visible/near-infrared and near-infrared/middle-infrared broad bands (≈ 100 nm width). Vegetation indices are particular combinations of the reflectance generally observed in two wavebands that aim to minimize the effect of confounding factors, while retaining a great sensitivity to canopy characteristics (Baret and Guyot 1991). Figure 12.15 shows the principles of this approach when applied using the normalized difference vegetation index (NDVI) computed with either visible/near-infrared bands or middle-infrared/near-infrared bands. Structural effects for a given leaf chlorophyll content lie along a single structural line. Low leaf area indices are close to the bare soil point where all the lines cross. Variation in chlorophyll content results in a change in the slope of the lines. A decrease in chlorophyll content corresponds to an increase in the slope of the structural line. It appears that for low leaf area indices, the discrimination between different levels of chlorosis will be difficult. Maximum separability is observed for full vegetation coverage. However, this approach is still sensitive to variations in

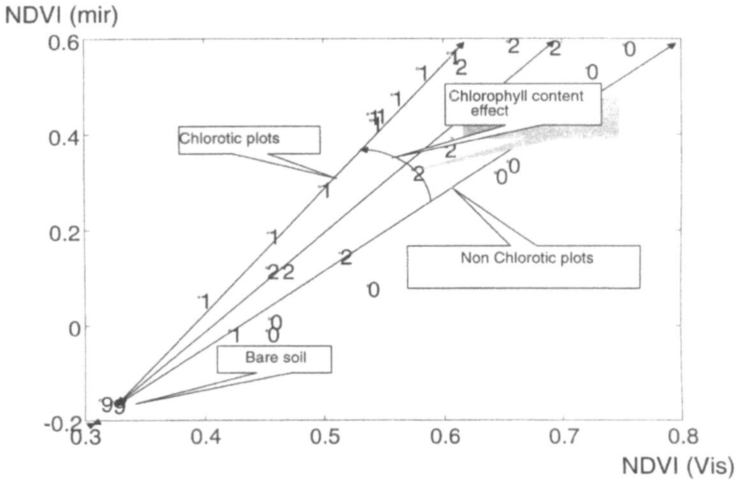

Fig. 12.15. Procedure proposed to estimate chlorophyll content from the combined use of the reflectance measured in the visible, near and middle infrared. Each plot is described with the normalized difference vegetation index (NDVI) using either a visible r_{vir} or middle infrared r_{nir} band combined with the near infrared bandρ_{nir}: NDVI (vis) = $(\rho_{nir} - \rho_{vist})/(\rho_{nir} + \rho_{vis})$: NDVI (mir) = $(\rho_{nir} - \rho_{mir})/(\rho_{nir} + \rho_{mirt})$. These experimental data were acquired in Grignon (France) in 1990 with sugar beet of varied leaf area index (Malthus 1990). Each *data point* corresponds to a particular plot. The chlorosis level corresponds to a given leaf chlorophyll content and is identified by a label 0, 1, or 2. *0* Nonchlorotic plots (Cab = 33 $\mu g\,cm^{-2}$); *1* highly chlorotic plots (Cab = 13 $\mu g\,cm^{-2}$); *2* intermediate plots (Cab = 23 $\mu g\,cm^{-2}$). Label *9* corresponds to the bare soil

bare soil reflectance because the vegetation indices used do not correct for all the soil background effects. An alternative approach has been proposed by Baret (1994) using neural networks.

Neural networks are recognized as a very powerful tool to discriminate between variables or to relate one set of variables to another. They are defined mainly by the type of neurons used, the way they are organized and connected (the network architecture), and the learning rule. A backpropagation network, which is one of the most commonly used networks (Rummelhart et al. 1986), was used in the study considered here (Baret 1994). Backpropagation networks are able to "learn" how to relate input variables to output variables. The learning is an iterative process during which the weights of the synaptic connections are adjusted to minimize the distance between the actual and the simulated output variables. In this study, we used a two-layer architecture, as shown in Fig. 12.16. The first layer is made of neurons with a sigmoidal transfer function. The second layer has only one neuron with a linear transfer function. The learning process is based on the learning data set which was independent of the test data set on which the network performances were evaluated.

The neural network approach was applied to a set of sugar beet plots with concurrent variations in soil background properties, leaf chlorophyll content (induced by herbicide applications), and leaf area index. The reflectance was observed in four wavebands to estimate leaf chlorophyll content: green (500–590 nm), red (620–680 nm), near-infrared (790–890 nm), and middle-infrared (1550–1750 nm). The results (Fig. 12.17) showed that the estimated chlorophyll contents were not in close agreement with the measured ones ($R^2 = 0.5391$). The error increased for low leaf area index values. This could easily be explained by the very low contribution of leaves to the radiometric signal under these conditions and, therefore, other confounding factors, such as variation in soil optical properties, will play a major role. This confirms the previous findings on broadband vegetation indices.

To encompass these limitations, we will investigate the possibility of estimating the canopy level chlorophyll content, which is the product between leaf chlorophyll content and leaf area index. We have used the same network architecture to relate canopy chlorophyll concentration to the reflectances observed in the four wavebands as described previously for leaf chlorophyll content. The estimation of canopy chlorophyll content was quite accurate as indicated by the $R^2 = 0.92$ value. The scatter around the 1 : 1 line increased slightly with canopy chlorophyll content. This was presumably due to the saturation effect, i.e., the lower sensitivity of reflectance to changes in l for large l values. Furthermore, the outlier observed for the test data set shows a large underestimation of canopy chlorophyll concentration. It corresponds to large, but not complete, vegetation cover (88%) and to the sand background. The increase in reflectance induced by this bright soil background could have been interpreted by the network as a decrease in both the chlorophyll concentration and leaf area index in the visible wavebands.

These empirical approaches may be limited to the conditions that prevailed during the experiments. For this reason, we also investigated the use of a more analytical approach: the inversion of radiative transfer models.

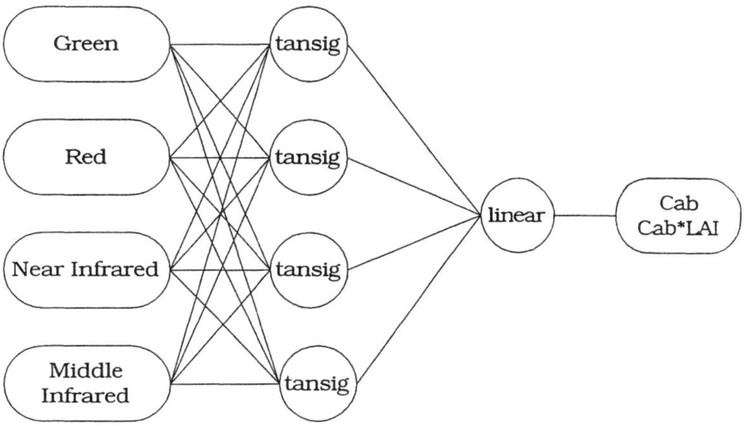

Fig. 12.16. Neural network architecture used. Input and output variables are represented by *boxes*. Neurons are represented by *circles*. *Circles tansig* and *linear* correspond to sigmoid and linear transfer functions, respectively. Each connection is represented by a *solid line* that corresponds to a given synaptic weight coefficient. A bias coefficient is also associated with each neuron

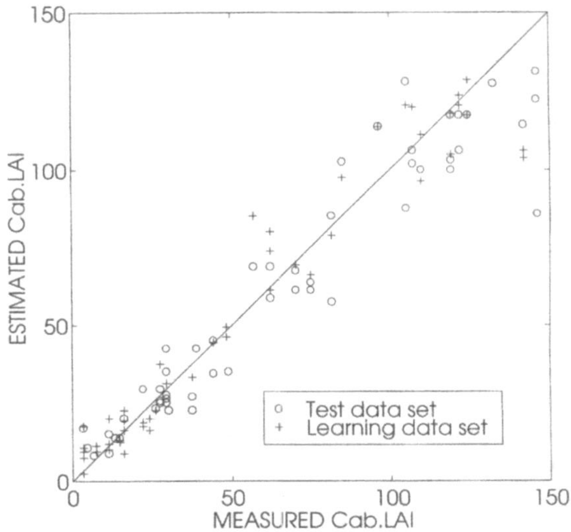

Fig. 12.17. Relationship between measured and estimated canopy chlorophyll content per unit area ($\mu g\,cm^{-2}$)

12.3.3 Model Inversion to Retrieve Leaf Chlorophyll Content

Canopy radiative transfer models allow us to compute canopy reflectance as a function of the input variables defined earlier, such as canopy structure and leaf chlorophyll content. The inversion of the radiative transfer model consists of finding the best set of input variables that minimizes the difference between the measured reflectance and the simulated ones. The minimization process is generally iterative and quite demanding of computer resources.

Jacquemoud (1993) demonstrated from a theoretical study that radiative transfer models were invertible from the spectral variation of canopy reflectance. He also showed that leaf chlorophyll content was estimated more reliably than canopy structural variables when the inversion was not too much constrained. Later, model inversion applied to experimental data was investigated over the same sugar beet data set containing large variations in leaf chlorophyll content, leaf area index, and soil background properties (Jacquemoud et al. 1995). The results showed that leaf chlorophyll content could be estimated with a reasonable accuracy (RMSE < $10\,\mu g\,cm^{-2}$) using either the 188 narrow bands (around 10 nm width) of the Airborne Visible and Infra Red Imaging Spectrometer (Green et al. 1992) or the six broad bands (around 100 nm width) of the satellite LANDSAT Thematic Mapper (Markham and Barker 1985). Figure 12.18 shows that the error around the 1:1 line is still significant, with an underestimation of C_{ab} when using the six TM bands as compared to the 188 AVIRIS bands. The inversion was constrained in this case by setting soil optical properties to their actual values. This constraint was imposed because of the lack of reliable models of the spectral variation of soil reflectance. When this constraint is not used, the canopy radiative transfer model actually gives very poor results. However, radiative transfer model inversion is quite new for vegetation applications. Presumably great improvements can be expected from:

- The optimisation of the sampling strategy, i.e., choice of wavebands, choice of measurement configuration.
- The refinement of leaf, canopy, and soil radiative transfer models.
- The design of enhanced inverse algorithms.

12.4 Conclusion

This chapter has shown the difficulty in developing general and reliable relationships that could be used to estimate directly the nitrogen concentration from measurements of fresh leaf reflectance or transmittance. This is mainly due to several factors:

- Nitrogen is not absorbing in itself, but through its bonds with light atoms such as C, H, or O. Therefore, it should be more easy to detect protein directly than

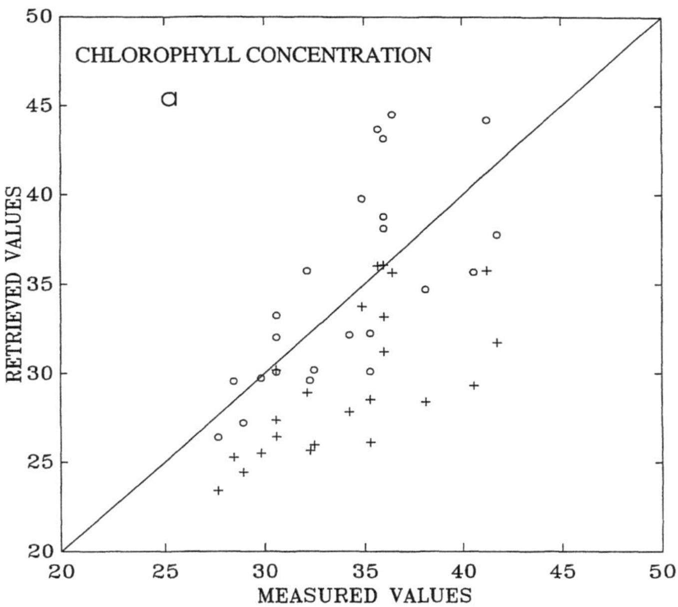

Fig. 12.18. Relationship between measured chlorophyll content per unit leaf area and that estimated using radiative transfer model inversion (μg cm^{-2}). Crosses and circles correspond respectively to the inversion performed with the 6 TM and the 188 AVIRIS bands

nitrogen, which might not be a major problem because of the generally strong relationship between nitrogen and protein concentration.
- Reflectance and transmittance are not directly sensitive to the concentration of absorbers, but to the content of absorber per unit leaf area. Thus, the specific leaf weight (SLW) should be used to convert from content to concentration.
- Protein is not present in large amounts in leaves, and the absorption features of protein are weak and not very specific between species.
- There exist no general relationships between other strong absorbers such as chlorophyll content or leaf dry mass content (SLW) and protein (or nitrogen) content.

Thus, estimation of nitrogen content or concentration must be established from empirical relationships with chlorophyll or SLW developed and applied in well-defined experimental conditions. Great attention should be paid to frequent verification of the applicability of the relations developed. However, regarding the experimental conditions that generally prevailed in a given area and for a given species and cultivar, it should be possible to derive such relationships. They could effectively complement the traditional sampling techniques by allowing an easier representation within the field or amongst fields. Improvements in the design of a device built to measure leaf chlorophyll content and possibly SLW may contribute to provide more generality to the relationships developed.

When transferring these conclusions to the reflectance observed at the canopy level, the structural effects and soil optical properties should be accounted for. The several approaches reviewed in this chapter show a relatively poor accuracy in determinations of the leaf chlorophyll content derived from canopy reflectance. This is due to the confounding effect of canopy structure, which is very important, particularly for small vegetation cover fractions. One way to increase the amount of vegetation viewed by the sensor would be to make oblique observations. Improvements of these techniques are expected, at least at the canopy level, allowing us to address afterwards the problem at the satellite level, which should include correction of atmospheric effects. The nitrogen status of canopies is generally represented by both a decrease in chlorophyll content and a reduction of the leaf area index. The combination of these two effects allows the development and evaluation of empirical relationships that should again be established in well-defined conditions. This is why the canopy level chlorophyll content at a given phenological stage is a potential variable to be investigated as a candidate indicator of plant nitrogen status. Remote sensing techniques at canopy or space levels should be seen as techniques complementary to the classical ones, allowing an extensive mapping of the results found in a limited sample of fields.

References

Badhwar GD (1984) Automatic corn-soybean classification using Landsat MSS data. II Early season crop proportion estimation. Remote Sens Environ 14:31,37

Baret F (1994) Use of reflectance spectral variation to retrieve canopy biophysical characteristics. In: Danson FM, Plummer SE (eds) Advances in environmental remote sensing. John Wiley, Chichester, pp 33–51

Baret F, Andrieu B (1994) Interêt du moyen infrarouge pour caracteriser la végétation. Bull Soc Fr Photogramm Télédétect 136(4):8–22

Baret F, Fourty T (1996) Spectral estimates of leaf water content and specific weight. Remote Sens Environ (Submitted)

Baret F, Guyot G (1991) Potentials and limits of vegetation indices for LAI and APAR assessment. Remote Sens Environ 35(2–3):161–173

Baret F, Champion I, Guyot G, Podaire A (1987) Monitoring wheat canopies with high spectral resolution radiometer. Remote Sens Environ 22(3):367–378

Baret F, Andrieu B, Guyot G (1988) A simple model for leaf optical properties in the visible and near infrared: application to the analysis of spectral shifts determinism. In: Lichtenthaler HK (ed) Applications of chlorophyll fluorescence. Kluwer, Dordrecht, pp 345–351

Baret F, Jacquemoud S, Guyot G, Leprieur C (1992) Modelled analysis of the biophysical nature of spectral shifts and comparison with information content of broad bands. Remote Sens Environ 41:133–142

Benedict HM, Swilder R (1961) Non destructive method for estimating chlorophyll content of leaves. Science 133:2015–2016

Boochs F, Kupfer G, Dockter K, Kubaush K (1990) Shape of the red edge as vitality indicator for plants. Int J Remote Sens 11(10):1741,1753

Breece HT, Holmes RA (1971) Bidirectional scattering characteristics of healthy green soybeans and corn leaves in vivo. Appl Optics 10(1):119–127

Chang SH, Collins W (1983) Confirmation of the airborne biogeophysical mineral exploration technique using laboratory methods. Econ Geol 78:723–736

Collins W, Chang SH, Raines G, Canney F, Ashley R (1983) Airborne biogeophysical mapping of hidden mineral deposits. Econ Geol 78:737–749

Curcio JA, Petty CC (1951) The near infrared absorption spectrum of liquid water. J Opt Soc Am 41(5):302–304

Curran PJ (1989) Remote sensing of foliar chemistry. Remote Sens Environ 30:271–278

Dardenne P (1990) Contribution àl'utilisation de la spectrométrie dans l'infrarouge pour l'étude de critères de qualité des céréales et des fourrages. Faculté des Sciences Agronomiques, Gembloux, pp 1–173

Dardenne P, Sinnaeve G, Biston R, Lecomte P (1991) Evaluation of NIT for predicting fresh forage quality. 4th Int NIRS Conf, Aberdeen

Demetriades-Shah TH, Steven MD (1988) High spectral resolution indices for monitoring crop growth and chlorosis. 4th Int Colloq Spectral signatures of objects in remote sensing, 18–22 Jan, Aussois, ESA SP-287

Demetriades-Shah TH, Steven MD, Clark JA (1990) High resolution derivative spectra in remote sensing. Remote Sens Environ 33:55–64

de Rosny G, Vanderhaghen R, Baret F, Equer B, Frangi JP (1995) A device for in situ measurements of leaf chlorophyll and carotenoid concentrations. In: Guyot G (ed) Photosynthesis and remote sensing. EARSel, Paris

Ercoli L, Mariotti M, Masoni A, Massantini F (1993) Relationship between nitrogen and chlorophyll content and spectral properties in maize leaves. Eur J Agron 2(2):113–117

Evans JR (1989) Photosynthesis and nitrogen relationships in leaves of C_3 plants. Oecologia 78: 9–19

Field C, Mooney HA (1986) The photosynthesis-nitrogen relationship in wild plants. In: Givnich TV (ed) On the economy of plant form and function. Cambridge University Press, New York, pp 25–55

Fourty T, Baret F (1996) Biochemical composition estimates from fresh leaf near infrared transmittance or reflectance spectra using the stepwise regression. Effect of the instrumental noise. Remote Sens Environ (Submitted)

Fourty T, Baret F, Jacquemoud S, Schmuck G, Verdebout J (1996) Leaf optical properties with explicite

description of its biochemical composition: direct and inverse problems. Remote Sens Environ 56:104–117
Green RO, Conel JE, Bregge CJ, Margolis JS, Carrere V, Vane G, Hoover G (1992) In-Flight calibration of the spectral and radiometric characteristics of AVIRIS in 1991. 3rd Annu JPL Airborne Geoscience Worksh, 1–2 June, Pasadena
Grossman YL, Ustin SL, Jacquemoud S, Sanderson E, Schmuck G, Verdebout J (1995) Critique of stepwise linear regression for the extraction of leaf biochemistry information from leaf reflectance data. Remote Sens Environ 56:182–193
Hardacre AK, Nicolson HF, Boyce MLP (1984) A portable photometer for the measurement of chlorophyll in intact leaves. N Z J Exp Agric 12:357–362
Hardwick K, Baker NR (1973) In vivo measurement of Chlorophyll content of leaves. New Phytol 72:51–54
Hosgood B, Jacquemoud S, Andreoli G, Verdebout J, Pedrini G, Schmuck G (1995) Leaf optical properties experiment 93 (LOPEX93). European Commission, Joint Research Center, Institute of Remote Sensing Applications, Ispra, Italy
Hruschka WR (1987) Data analysis: wavelength selection methods. In: Williams PC, Norris KH (eds) *Near infrared technology in the agricultural and food industries.* Chap 3, American Association of Cereal Chemists, St Paul
Jacquemoud S (1993) Inversion of the PROSPECT + SAIL canopy reflectance model from AVIRIS equivalent spectra: theoretical study. Remote Sens Environ 44:281–292
Jacquemoud S, Baret F (1990) PROSPECT: a model of leaf optical properties spectra. Remote Sens Environ 34:75–91
Jacquemoud S, Baret F, Andrieu B, Danson M, Jaggard K (1995) Extraction of vegetation biophysical parameters by inversion of the PROSPECT + SAIL model on sugar beet canopy reflectance data. Application to TM and AVIRIS sensors. Remote Sens Environ 52:163–172
Macnicol PK, Dudzinski ML, Condon BN (1976) Estimation of chlorophyll in tobacco leaves by direct photometry. Ann Bot 40:143–152
Malthus TJ (1990) Anglo-French collaborative reflectance experiment. Experiment I, Broom's Barn Experimental Station, July 1989. INRA bioclimatologie, BP 91, 84143 Montfavet France
Markham BL, Barker JL (1985) Spectral characteristics of the LANDSAT Thematic Mapper sensors. Int J Remote Sens 6(5):597–716
Marten GC, Shenk JS, Barton FE (1989) Near infrared reflectance spectroscopy (NIRS): analysis of forage quality, vol 643. *US Department of Agriculture Handbook*, Washington, DC, pp 1–96
Miller JR, Hare EW, Neville RA, Gauthier RP, McColl D, Till SM (1985) Correlation of metal concentration with anomalies in narrow band multispectral imagery of the vegetation red reflectance edge. 4th Int Symp Remote sensing of environment, 1–4 April, San Francisco
Miller JR, Hare EW, Wu J (1990) Quantitative characterization of the vegetation red edge reflectance. 1. An inverted-gaussian model. Int J Remote Sens 11(10):1755–1773
Milton NM, Mouat DA (1989) Remote sensing of vegetation responses to natural and cultural environmental conditions. Photogramm Engin Remote Sens 55(8):1167–1173
Minolta Camera Co (1991) Manual for chlorophyll meter SPAD-502. Minolta Camera, Osake
Peng S, Garcia FV, Laza RC, Cassman KG (1993) Adjustment for specific leaf weight improves chlorophyll meter's estimate of rice leaf nitrogen concentration. Agron J 85:987–990
Penuelas J, Filella I, Baret F (1994) Semi-empirical indices to assess carotenoids/chlorophyll ratio from leaf spectral reflectance. Photosynthetica 31(2):221–230
Rock BN, Miller JR, Moss DM, Freemantle JR, Boyer MJ (1990) Spectral characterization of forest damage occurring on Whiteface Mountain (NY) – studies with the fluorescence line imager (FLI) and ground-based spectrometers. In: Proc SPIE Conf, Technical Symp on Optical engineering and photonics in aerospace sensing, 16–20 April, Orlando. Imaging spectroscopy of the terrestrial environment, SPIE 1298, pp 190–201
Rummelhart DE, Hinton GE, Williams RJ (1986) Learning internal representations by error propagation. In: Rummelhart D, Mc Clelland J (eds) Parallel data processing, vol 1. MIT Press, Cambridge, pp 318–362

Spomer LA, Smith MAL, Sawwan JS (1988) Rapid nondestructive measurement of chlorophyll content in leaves with nonuniform chlorophyll distribution. Photosynth Res 16:277–284

Turner FT, Jund MF (1991) Chlorophyll meter to predict nitrogen topdress requirement for semidwarf rice. Agron J 83:926–928

Verhoef W (1984) Light scattering by leaf layers with application to canopy reflectance modeling: the SAIL model. Remote Sens Environ 16:125–141

Verhoef W (1985) Earth observation modeling based on layer scattering matrices. Remote Sens Environ 17:165–178

Wallihan EF (1973) Portable reflectance meter for estimating chlorophyll concentrations in leaves. Agron J 65:659–662

Williams PC, Norris KH (1987) Near infrared technology in agricultural and food industries. American Association of Cereal Chemists, St Paul, 330 pp

Yadava VL (1986) A rapid non-destructive method to determine chlorophyll in intact leaves. Hortic Sci 21(6):1449–1450

CHAPTER 13

Concluding Remarks: N Hazards to Crops and Environment

B. Mary

Contents

13.1 Towards a Better Adjustment of N Fertilization

The arguments given in this book have indicated that the concept of critical nitrogen concentration in plants has a strong theoretical and experimental basis. The critical N% curve versus cumulative dry matter appears to be largely independent of environmental cropping conditions and plant genotypes. It is even rather independent of the crop species, at least if we consider separately plants with a C_3 or C_4 metabolism although some small variations within each of these two groups cannot be excluded. It is a unifying concept which can be applied to all arable crops, grasslands, and possibly to forest trees.

The critical N% curve therefore constitutes an objective criterion for defining "crop nitrogen requirements" and a reference method for making a diagnosis of N nutrition during crop growth. Both properties are needed for improving current N-fertilizer recommendations. Firstly, the determination of crop N requirements is essential in N models, particularly those used for practical N recommendations which can be either "static" balance-sheet models such as the "Nmin method" (Wehrmann and Scharpf 1989) or the more detailed AZOBIL method (Machet et al. 1990), or dynamic models such as SUNDIAL (Smith et al. 1996). The accurate prediction of crop N requirements versus time is necessary to achieve the objectives in both quantity and quality of harvested products. If yield losses due to N excess like lodging can be overcome using growth regulators, the negative effects

INRA Unité d'Agronomie, rue Fernand Christ, 02007 Laon Cédex, France

Diagnosis of the Nitrogen Status in Crops
G. Lemaire (ed)

of N excess on crop quality cannot be avoided: this is the case, for example, with malting barley whose grain protein content must be low (see Chap. 4) and sugar beet for which accumulation of amino-N is harmful to sugar extraction or vegetable crops which can accumulate undesirable nitrate. Conversely, some crops like soft or durum wheat must produce grain with a high protein content: they may require to be conducted above the critical N% curve, at least temporarily.

Secondly, the possibility of making a reliable N nutrition diagnosis during crop growth is particularly promising in order to update and eventually modify the fertilization rates proposed by these models. The JUBIL method, an example of this strategy, combines a predictive soil indicator (mineral N balance) and a real-time plant indicator (nitrate concentration in the stem base). The advantage of this strategy is to correct possible deviations of model predictions which may be due to either model imperfections or weather hazards. Undoubtedly, these types of methods will be used more and more in the near future if they remain easy to handle and cheap enough to be profitable to farmers. This book also shows the future potential application of various radiometric methods in order to improve crop N diagnosis or to recalibrate simulation models during crop growth. Such physical measurements could be made either on destructive plant samples or, more interestingly, on undisturbed crop canopy using remote sensing techniques. However, these techniques have to be investigated in greater detail before they can be applied.

13.2
Towards Reduced Losses of N in the Environment

However, the objective of agriculture is not only to be productive but also to be "sustainable", i.e. to preserve the soil, air and water quality in the long term. Undoubtedly, intensive agriculture has led to an increase in N leakage in the environment, in the atmosphere as ammonia and nitrogen oxides, and in water resources as nitrate. The challenge is to define the cropping techniques which will give the best compromise between good agricultural production (in quantity and quality) and an acceptable level of N losses.

It is clear that a better adjustment of N fertilization can only be beneficial in reducing N losses, if a comparison is made to situations in which N fertilizer is applied in excess. From that point of view, a substantial reduction in nitrate leaching and denitrification can be expected from the development of the previously mentioned methods of fertilization, since both processes are stimulated by an increase in nitrate concentration in soils (Hénault and Germon 1995).

However, the importance of the reduction in N losses due only to improved N fertilization is still under debate. If we consider the nitrate pollution problem, the objective is to minimize the mean concentration of water percolating to the aquifers and to match the EU limit (50 mg NO_3/l). There is still ongoing discussion between those who consider that a better N adjustment and a fortiori a reduction in N fertilizer inputs can be efficient for solving the problem (e.g., Davies and Sylvester-Bradley 1995) and those who think like Macdonald et al. (1989) that "even a drastic reduction in N fertiliser use would have little effect on nitrate leaching". In fact, it is necessary to distinguish between annual and long-term effects of N fertilizers.

13.2.1
Annual Effects

13.2.1.1
Residual N at Harvest

It is now well demonstrated that the amounts of mineral nitrogen found in the soil at crop harvest (residual N) are not proportional to the amount of fertilizer applied to the crop. Figure 13.1 illustrates the mean results obtained by Chaney (1990) which have been confirmed by other studies (Guiot and Grevy 1990; Glendining et al. 1992; Recous et al. 1996). It shows the evolution of grain yield and residual N at harvest versus the amount of fertilizer N applied in wheat experiments made in England. The residual mineral N in soil remained almost constant independently of the N rate as long as the maximum yield was not obtained. It then increased when the N rate exceeded the optimum rate, but at a lower rate than the excess of fertilizer N. For a moderate excess of fertilizer (0–100 kg N/ha), the average increase in residual N is about 40–50% of the fertilizer-N excess. There are three main reasons explaining that the effects of overfertilization are alleviated. Their importance can be ranked in decreasing order:

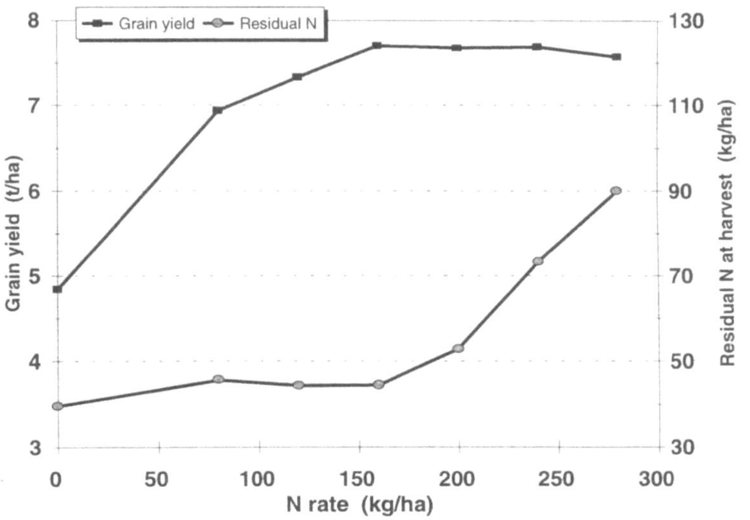

Fig. 13.1. Evolution of grain yield and mineral N in soil at harvest (0–90 cm) versus fertilizer N rate applied to winter wheat crops (data from Chaney 1990). Mean of seven experiments (excluding Aynho 88)

1. The plant N uptake still increases after the optimum rate, even though the maximum yield or shoot biomass has been reached: previous results (see Chap. 4) have indicated that the N uptake above the critical N curve can be very important.
2. The microbial immobilization of fertilizer-N increases with the N rate, at least to a certain extent. This is suggested by the results of various ^{15}N experiments showing that the ^{15}N recovered in organic form at harvest was rather independent on the amount of fertilizer-N added.
3. The gaseous losses – volatilization or denitrification – are favored by high NH_4^+ or NO_3^- concentrations in soil (Hénault and Germon 1995; Génermont 1996).

In spite in this "buffer effect", the interest of fertilizing at the optimum to reduce the residual N remaining in soil at harvest is clear. Another interesting example drawn from the work of Justes (1993) is shown in Fig. 13.2. It presents the average residual N found at harvest for various N rates applied in six wheat experiments conducted in France. X was the rate recommended by the AZOBIL method, based on soil measurements made in February (during tillage). The JUBIL method recommended the application of only (X-40) kg/ha in half of the experiments and the application of (X-40) + 40 in the other half. The comparison with the other N rates tested indicated that this strategy gave the optimum yield in all experiments. The optimum fertilization rate was then on average X-20 kg/ha. Figure 13.2 shows that this strategy minimized the residual N in soil at harvest; an extra and significant 10 kg N/ha was found in the experiments receiving X (recommended rate without plant indicator) and an extra 22 kg N/ha in the plots receiving X + 40 (which can be considered a "safety practice" for farmers). The extra 10 kg/ha might appear small,

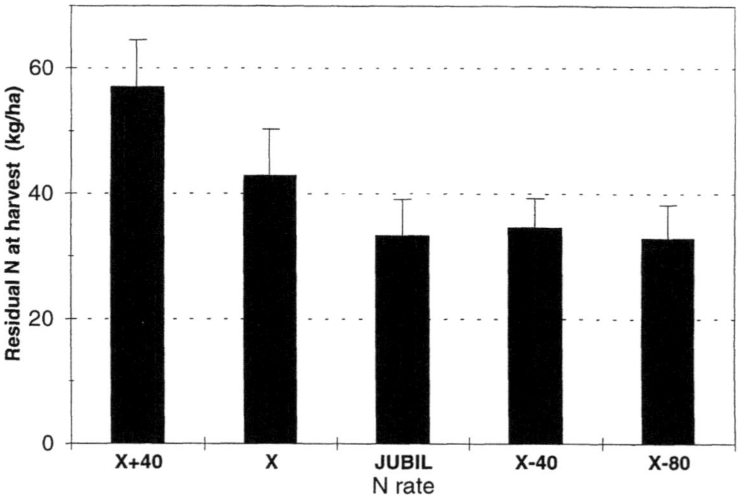

Fig. 13.2. Evolution of mineral N in soil at harvest (0–90 cm) versus fertilizer N rate applied to winter wheat crops (data from Justes 1993). Mean of six experiments. *X* represents the optimum rate calculated by the AZOBIL method. The rate determined by the JUBIL method corresponded either to (X-40) kg/ha (two dressings) or to (X-40) + 40 kg/ha (three dressings)

but the nitrate present in the soil at harvest is the first nitrate to be leached out when drainage starts again. If we consider a mean annual drainage of 220 mm, saving 10 kg N/ha from leaching reduces the concentration of drained water by about 20 mg NO_3/l, which is very significant. A larger benefit can even be expected under farming conditions where "safety practices" are thought to be frequent.

13.2.1.2
N Mineralized During the Following Autumn and Winter

The effect of overfertilization can also appear after crop harvest, during autumn and winter, by stimulating soil N mineralization. ^{15}N experiments have indicated that the absolute amount of N mineralized derived from the fertilizer (remineralization of N immobilized during crop growth) was much smaller than the N derived from the soil, even though the N derived from the fertilizer was proportionally more mineralizable than the "native" soil organic N (Shen et al. 1989). Results obtained in lysimeters (Denys et al. 1990) confirm this conclusion as well as our own ^{15}N experiments (Laurent and Mary 1992; Recous et al. 1996). Figure 13.3 shows the importance of net mineralization during an intercropping period between wheat harvest and sowing of sugar beet: the greater part (more than 80 kg N/ha) was derived from soil, the N derived from the fertilizer being only 7 kg N/ha. This type of result led Addiscott et al. (1991) to conclude that "this

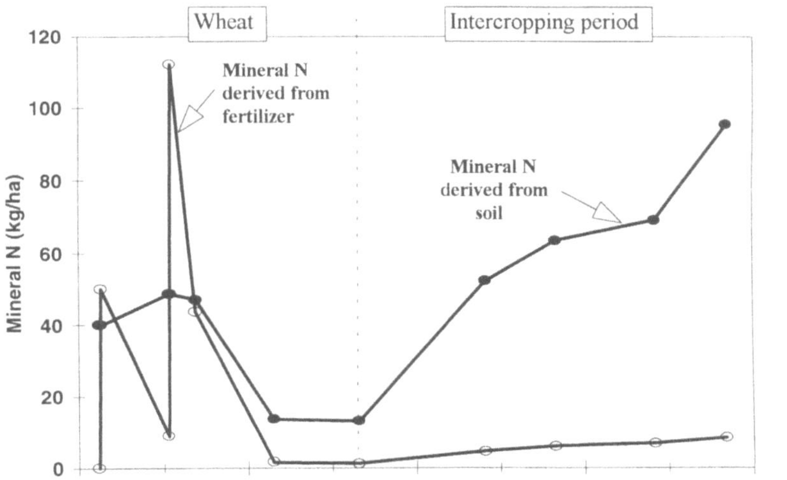

Fig. 13.3. Evolution of mineral N in soil (0–120 cm) during a winter wheat-sugar beet rotation in France (Laurent and Mary 1992). The wheat crop received two applications of ^{15}N-labeled fertilizer (50 and 110 kg/ha), which allowed one to distinguish N derived from fertilizer and N derived from "native" soil N

naturally occurring nitrate is the biggest cause of the nitrate problem, responsible for a far greater proportion of the nitrate problem than direct losses from nitrogen fertiliser".

13.2.2
Long-Term Effects

We believe that this conclusion must be considered with care because the small contributions of N derived from fertilizer each year may become significant over time due to cumulative effects. An important and possibly major effect of fertilizer on nitrate pollution could be a long-term effect due to a more and more positive N balance with increasing N inputs. Davies and Sylvester-Bradley (1995) have estimated that the average N balance for UK winter wheat has shifted from-27 kg/ha in 1957 to +18 kg/ha in 1992. The long-term effect would then be an increase in annual net mineralization which can have two origins: (1) a straightforward buildup in soil organic matter; (2) an increase in the rate of mineralization per unit of organic N in soil.

The first possibility cannot be excluded, although it seems that the organic matter level has decreased or stabilized in many arable soils during the last 40 years. The second hypothesis is strongly suggested by the results of Jenkinson's group obtained in the oldest, long-term experiment in the world (Johnston and Jenkinson 1989). Figure 13.4 shows that the amounts of organic N in soil increased moderately with the N fertilizer rate applied, at least between 0 and 144

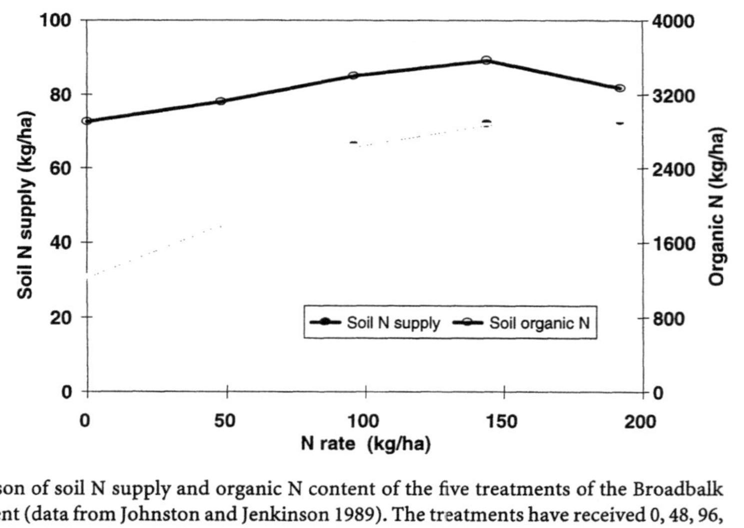

Fig. 13.4. Comparison of soil N supply and organic N content of the five treatments of the Broadbalk long-term experiment (data from Johnston and Jenkinson 1989). The treatments have received 0, 48, 96, 144 kg N/ha since 1852 or 192 kg N/ha since 1968. Soil N supply to wheat crop was calculated for 1 year by labeling the fertilizers

kg ha^{-1} $year^{-1}$. The soil N supply, estimated by the wheat recovery of unlabeled N after adding labeled fertilizer, was markedly more influenced by the N applied than total N. It represented 1.1% of the soil organic N content in the 0 fertilizer treatment and 2.0% in the treatment receiving 144 kg/ha each year since 1852. The same trend was found, but less clearly, with two other indicators of soil N availability: N mineralized in laboratory incubation and microbial biomass size.

This discussion shows that the expected impact of improved application of N fertilization on N losses in the environment is not known precisely as yet. We believe that the direct effects of such measures on nitrate leaching are probably greater than those expected from the previously mentioned studies, because the diversity of agricultural situations has not been fully explored in these studies. Other crop conditions (rapeseed, potatoes, vegetable crops, etc.) and soils (shallow or sandy soils) are more susceptible to fertilizer N losses. The improvement of fertilization recommendations and applications must therefore be strongly encouraged, particularly because the importance of their effect on water quality will appear in years or decades.

References

Addiscott TM, Whitmore AP, Powlson DS (1991) Sources of nitrate leakage: arable farming. In: Farming, fertilizers and the nitrate problem. CAB International, Wallingford, pp 92–109

Chaney K (1990) Effect of nitrogen fertilizer rate on soil nitrogen content after harvesting winter wheat. J Agric Sci 114:171–176

Davies DB, Sylvester-Bradley R (1995) The contribution of fertiliser nitrogen to leachable nitrogen in the UK. J Sci Food Agric 68:399–406

Denys D, Muller JC, Mariotti A (1990) Consequences of the organization of the mineral nitrogen from a fertilizer on bioavailability and lixiviation. In: Calvet R (ed) Nitrates-agriculture-eau. INRA Editions, Paris, pp 189–194

Génermont S (1996) Modélisation de la volatilisation d'ammoniac après épandage de lisier sur parcelle agricole. Thèse Docteur Université Paul Sabatier, Toulouse, 278 pp + annexes

Glendining MJ, Poulton PR, Powlson DS (1992) The relationship between inorganic N in soil and the rate of fertilizer N applied on the Broadbalk wheat experiment. Aspects Appl Biol 30:95–102

Guiot J, Grevy L (1990) Evolution des nitrates dans une terre soumise à la rotation betterave-froment-escourgeon. In: Calvet R (ed) Nitrate-agriculture-eau. INRA Editions, Paris, pp 417–423

Hénault C, Germon JC (1995) Quantification of denitrification and nitrous oxide production in soils. Agronomie 15:321–355

Johnston AE, Jenkinson DS (1989) The nitrogen cycle in UK arable agriculture. Proc No 286, The Fertiliser Society of London, pp 3–24

Justes E (1993) Diagnostic de la nutrition azotée du blé à partir de la teneur en nitrate de la base de la tige. Thèse Docteur INA-PG, Paris, 227 pp + annexes

Laurent F, Mary B (1992) Management of nitrogen in farming systems and the prevention of nitrate leaching. Aspects Appl Biol 30:45–61

Macdonald AJ, Powlson DS, Poulton PR, Jenkinson DS (1989) Unused fertiliser nitrogen in arable soils – its contribution to nitrate leaching. J Sci Food Agric 46:407–419

Machet JM, Dubrulle P, Louis P (1990) Azobil, a computer program for fertilizer N recommendations based on a predictive balance-sheet method. Proc the 1st ESA Congr, Paris, pp 2–21

Recous S, Jeuffroy MH, Mary B, Meynard JM (1996) Gestion de l'azote en zone d'agriculture intensive. Rapport de contrat SCGP. Laon, 37 pp

Shen SM, Hart PBS, Powlson DS, Jenkinson DS (1989) The nitrogen cycle in the Broadbalk wheat experiment: ^{15}N-labelled fertilizer residues in the soil and in the soil microbial biomass. Soil Biol Biochem 21:529–533

Smith JU, Bradbury NJ, Addiscott TM (1996) Sundial: a PC based system for simulated nitrogen dynamics in arable land. Agron J 88:38–43

Wehrmann J, Scharpf HC (1989) Reduction of nitrate leaching in a vegetable farm: fertilisation, crop rotation, plant residues. In: Germon JC (ed) Management systems to reduce impact of nitrates. Elsevier, London, pp 147–156

Subject Index

Springer and the environment

At Springer we firmly believe that an international science publisher has a special obligation to the environment, and our corporate policies consistently reflect this conviction.

We also expect our business partners – paper mills, printers, packaging manufacturers, etc. – to commit themselves to using materials and production processes that do not harm the environment. The paper in this book is made from low- or no-chlorine pulp and is acid free, in conformance with international standards for paper permanency.

MIX
Papier aus verantwortungsvollen Quellen
Paper from responsible sources
FSC® C105338

If you have any concerns about our products,
you can contact us on
ProductSafety@springernature.com

In case Publisher is established outside the EU,
the EU authorized representative is:
Springer Nature Customer Service Center GmbH
Europaplatz 3, 69115 Heidelberg, Germany

Printed by Libri Plureos GmbH
in Hamburg, Germany